Dorin Bucur – Giuseppe Buttazzo

Variational Methods
in Some Shape Optimization Problems

APPUNTI

SCUOLA NORMALE SUPERIORE
2002

ISBN: 978-88-7642-297-3

Contents

Preface

The study of shape optimization problems is a very wide filed, both classical, as the isoperimetric problem and the Newton problem of the best aerodynamical shape show, and modern, for all the recent results obtained in the last two, three decades.

The fascinating feature is that the competing objects are *shapes*, i.e. domains of $\mathbb{R}^N$, instead of functions, as it usually occurs in problems of the calculus of variations. This constraint often produces additional difficulties that lead to a lack of existence of a solution and the introduction of suitable *relaxed* formulations of the problem. However, in some few cases an optimal solution exists, due to the special form of the cost functional and to the geometrical restrictions on the class of competing domains.

This volume collects the lecture notes of two courses given in the academic year 2000-2001 by the authors at the Dipartimento di Matematica of University of Pisa and at Scuola Normale Superiore di Pisa respectively. The courses were mainly addressed to Ph. D. students and required as a background the topics in functional analysis that are usually taught in undergraduate courses.

The style these notes have been written is quite informal and follows the lectures that have been given; some lack of precision may then be present, together with several typos. The list of references is also far to be complete and exhaustive, both for our partial knowledge of all the papers published in this field as well as for the reason that the subject is still quickly developing. Any comment about the notes is welcome; please address them to the authors, respectively at `bucur@math.univ-fcomte.fr` and `buttazzo@dm.unipi.it`.

Pisa, January 20, 2002

D. Bucur and G. Buttazzo

Chapter 1

Introduction to shape optimization theory and some classical problems

In this chapter we introduce a shape optimization problem in a very general way and we discuss some of their features that will be considered in the following chapters. We also present some classical problems like the isoperimetric problem and some of its variants, which can be viewed in the framework of shape optimization.

A shape optimization problem is a minimization problem where the unknown variable runs over a class of domains; then every shape optimization problem can be written in the form

$$\min\left\{F(A) \ : \ A \in \mathcal{A}\right\}$$

where $\mathcal{A}$ is the class of admissible domains and F is the cost function that one has to minimize over $\mathcal{A}$.

It must be noticed that the class $\mathcal{A}$ of admissible domains does not have any linear or convex structure, so in shape optimization problems it is meaningless to speak of convex functionals and similar notions. Moreover, even if several topologies on families of domains are available, in general there is no an a priori choice of a topology in order to apply the direct method of the calculus of variations, for obtaining the existence of at least an optimal domain.

We want to stress that, as also it happens in other kinds of optimal control problems, in several situations an optimal domain does not exist;

this is mainly due to the fact that in these cases the minimizing sequences are highly oscillating and converge to a limit object only in a *"relaxed"* sense. Then we may have, in these cases, only the existence of a *"relaxed solution"* that in general is not a domain, and whose characterization may change from problem to problem.

We shall introduce in the next chapters a general procedure to relax optimal control problems and in particular shape optimization problems. A case which will be considered in detail is when a Dirichlet condition is imposed on the free boundary: we shall see that in general one should not expect the existence of an optimal solution. However, the existence of an optimal domain occurs in the following cases:

i) when severe geometrical constraints on the class of admissible domains are imposed (see Section 5.1);

ii) when the cost functional fulfills some particular qualitative assumptions (see Section 5.4);

iii) when the problem is of a very special type, involving only the eigenvalues of the Laplace operator, where neither geometrical constraints nor monotonicity of the cost are required (see Section 6.3).

Far to be an exhaustive classification, this is simply the state of the art at present.

The case when Neumann conditions are considered on the free boundary is not discussed in this Lecture notes. We refer the reader to papers [41], [47], [48], [78], [79], [101] for some topics related to this subject.

In this chapter we present some problems of shape optimization that can be found in the classical literature. In all cases we consider here, the existence of an optimal domain is due to the presence either of geometrical constraints in the class of admisible domains or of geometrical penalizations in the cost functional. The standard background of Functional Analysis and of function spaces (as Sobolev or *BV* spaces) is assumed to be known.

Among the classical questions which can be viewed as shape optimization problems we include the isoperimetric problem which will be presented in a great generality and with several variants. In order to set the problem correctly, the notion of perimeter of a set is required; this will be introduced by means of the theory of *BV* functions. The main properties of *BV* functions will be recalled and summarized without entering into details; the reader interested to finer results and deeper discussions will be referred to one of the several books available in the field (see for instance [8], [104], [12], [155], [126]).

Another classical question which can be considered as a shape optimization problem is the determination of the best aerodynamical profile for a body in a fluid stream under some constraints on its size. The Newton model for the aerodynamical resistance will be considered and various kinds of constraints on the body will be discussed.

The Newton problem of optimal aerodynamical profiles gives us the opportunity to consider in the next chapter a larger class of shape optimization problems: indeed we shall take those whose admissible domains are convex. This geometrical constraint allows in several cases to obtain the existence of an optimal solution.

In Section 1.4 we consider a problem of optimal interface between two given media; either a perimeter constraint on the interface, or a perimeter penalization, gives in this case enough compactness to guarantee the existence of an optimal classical solution. This problem gives us the opportunity to discuss some properties of Γ-convergence, which plays an important role in several shape optimization problems.

In section 1.5 we deal with the problem of finding the optimal shape of a thin insulating layer around a thermally conducting body. The problem will be set as an optimal control problem, where the thickness function of the layer will be the control variable and the temperature will be the state variable.

1.1 General formulation of a shape optimization problem

As already said above, a shape optimization problem is a minimization problem of the form

$$(1.1) \qquad\qquad \min\big\{F(A) \ : \ A \in \mathcal{A}\big\}$$

where $\mathcal{A}$ is the class of admissible domains and F is the cost functional. We shall see that, unless some geometrical constraints on the admissible sets are assumed or some very special cases for cost functionals are considered, in general the existence of an optimal domain may fail. In these situations the discussion will then be focused on the relaxed solutions, that always exist.

We shall see that, in order to give a qualitative description of the optimal solutions of a shape optimization problems, it is important to derive the so called *necessary conditions of optimality*. These conditions, as it usually happens in all optimization problems, have to be derived from the comparison of the cost of an optimal solution A with the cost of other suitable admissible choices, close enough to A. This procedure is what is usually called a *variation* near the solution. The difficulty in obtaining necessary conditions of optimality for shape optimization problems consists in the fact that, being the unknowns domains, the notion of neighbourhood is not a priori clear; the possibility of choosing a domain variation could then be rather wide. The same method can be applied, when no classical solution exists, to relaxed solutions, and this will provide qualitative information about the behaviour of minimizing sequences of the original problem.

Finally, for some particular problems presenting special behaviours or symmetries, one would like to exhibit explicit solutions (balls, ellipsoids, ...). This could be very difficult, even for simple problems, and often, instead of having established results, we can only give conjectures.

In general, since the explicit computations are difficult, one should develop efficient numerical schemes to produce approximated solutions; this is a challenging field we will not enter; we refer the interested reader to some recent books and papers (see for instance references [4, 123, 167, 175]).

1.2   The isoperimetric problem and some of its variants

The first and certainly most classical example of a shape optimization problem is the isoperimetric problem. It can be formulated in the following way: find, among all admissible domains with a given *perimeter* (this explains the term "isoperimetric"), the one whose Lebesgue measure is as large as possible. Equivalently, one could minimize the perimeter of a set among all admissible domains whose Lebesgue measure is prescribed.

The first difficulty consists in finding a definition of perimeter general enough to be applied to nonsmooth sets and to allow us to apply the direct method of the calculus of variations. The definition below goes back to De

Giorgi (see [104]) and is now considered classical; we assume the reader is familiar with the spaces of functions with bounded variations and with their properties.

Given a set $A \subset \mathbb{R}^N$ we denote by 1_A the characteristic function of A, defined by

(1.2)
$$1_A(x) = \begin{cases} 1 & \text{if } x \in A \\ 0 & \text{otherwise.} \end{cases}$$

Definition 1.2.1 *We say that a set A with finite Lebesgue measure is a set of finite perimeter in $\mathbb{R}^N$ if its characteristic function 1_A belongs to $BV(\mathbb{R}^N)$. This means that the distributional gradient $\nabla 1_A$ is a vector valued measure with finite total variation. The total variation $|\nabla 1_A|$ is called perimeter of A.*

The admissible domains A we consider are constrained to be contained in a given closed subset K of $\mathbb{R}^N$. Instead of fixing their Lebesgue measure, more generally we impose the constraint

$$\int_A f(x)\, dx = c$$

where c is a given constant and f in a given function in $L^1_{loc}(\mathbb{R}^N)$. Note that when f is a constant function the class of admissible domains is simply the class of all subsets of K with a given volume.

With this notation the isoperimetric problem can be then formulated in the following way:

given a closed subset K of $\mathbb{R}^N$ and a function $f \in L^1_{loc}(\mathbb{R}^N)$ find the subset of K whose perimeter is minimal, among all subsets A of K whose integral $\int_A f(x)\, dx$ is prescribed.

We have then a minimization problem of the form (1.1) with

$$F(A) = \text{Per}(A) = \int |\nabla 1_A|$$

$$\mathcal{A} = \left\{ A \subset K \ : \ \int_A f(x)\, dx = c \right\}.$$

Note that all subsets of K with infinite perimeter are ruled out by the formulation above, because the cost functional evaluated on them takes the value $+\infty$.

Theorem 1.2.2 *With the notation above, if K is compact and if the class of admissible sets is nonempty, then the minimization problem*

(1.3) $$\min\{F(A) \ : \ A \in \mathcal{A}\}$$

admits at least a solution.

Proof The proof follows the usual scheme of the direct methods of the calculus of variations. Take a minimizing sequence (A_n), the perimeters $\text{Per}(A_n)$ are then equi-bounded; since $A_n \subset K$ and since K is bounded the measures of A_n are equi-bounded as well. Therefore the sequence 1_{A_n} is bounded in $BV(Q)$ being Q a large ball containing K; we may then extract a subsequence (which we still denote by the same indices) which converges weakly* to some function $u \in BV(Q)$ in the sense that

$$\begin{cases} 1_{A_n} \to u \text{ strongly in } L^1(Q) \\ \nabla 1_{A_n} \to \nabla u \text{ weakly* in the sense of measures.} \end{cases}$$

The function u has to be of the form 1_A for some set A with finite perimeter. Moreover, we obtain easily that $A \subset K$ and $\int_A f(x)\,dx = c$, which shows that A is an admissible domain. This domain A achieves the minimum of the cost functional since (as it is well known) the perimeter is a weakly* lower semicontinuous function on BV.   ∎

Example 1.2.3 When K is not bounded, the existence of an optimal domain may fail. In fact, take $K = \mathbb{R}^N$ and $f(x) = |x|$. It is clear that a ball $B_{x_0,r}$ with $|x_0| \to +\infty$ and $r \to 0$ suitably chosen may fulfill the integral constraint; on the other hand the perimeter of a such $B_{x_0,r}$ goes to zero. Then the infimum of the problem is zero, which is clearly not attained.

Example 1.2.4 If K is unbounded, the existence of an optimal domain for problem (1.3) may fail even if $f \equiv 1$. In fact, let c be the measure of the unit ball in $\mathbb{R}^N$ and let

$$K = \bigcup_{n \in \mathbb{N}} B_{x_n,r_n}$$

where (r_n) is a strictly increasing sequence of positive numbers converging to 1 (for instance $r_n = 1 - 1/n$) and (x_n) is a sequence of points in $\mathbb{R}^N$ such that $|x_n - x_m| \geq 2$ if $n \neq m$. Then it is easy to see that the infimum of problem (1.3) is given by the value $\text{Per}(B_{0,1})$ which is not attained, since the set K does not contain any ball of radius 1 (see figure below).

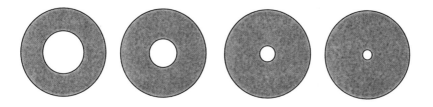

Figure 1.1: The set K.

The case $K = \mathbb{R}^N$ and $f \equiv 1$ is the classical isoperimetric problem. It is well known that in this case the optimal domains are the balls of measure c, even if the proof of this fact is not trivial. In the case $N = 2$ the proof can be obtained in an elementary way by using the *Steiner symmetrization* method; in higher dimensions the proof is more involved. It is not our goal to enter in this kinds of details and we refer the interested reader to the wide literature on the subject.

A variant of the isoperimetric problem consists in counting in the cost functional only the part of the boundary of A which is interior to K. More precisely, we consider an open subset D of $\mathbb{R}^N$ with a Lipschitz boundary and we define the perimeter relative to D of a subset A as

$$\text{Per}_D(A) = |\nabla 1_A|(D).$$

In this way a set A will be of finite perimeter in D if the function 1_A belongs to the space $BV(D)$.

Again, we have a minimization problem of the form (1.1) with

$$F_D(A) = \text{Per}_D(A) = \int_D |\nabla 1_A|$$

$$\mathcal{A} = \left\{ A \subset D \ : \ \int_A f(x)\, dx = c \right\}.$$

Theorem 1.2.5 *With the notation above, if D is bounded and if the class of admissible sets is nonempty, then the minimization problem*

(1.4) $$\min \left\{ F_D(A) \ : \ A \in \mathcal{A} \right\}$$

admits at least a solution.

Proof The proof can be obtained by repeating step by step the proof of Theorem 1.2.2. ∎

Example 1.2.6 If the assumption that D is bounded is dropped, it is easy to construct counterexamples to the existence result above, even if the datum f is identically equal to 1. In fact, if we define the function

$$\phi(x) = -8x + 8 \quad \text{if } x \in]1/2, 1]$$
$$\phi(x) = -32x + 16 \quad \text{if } x \in]1/4, 1/2]$$
......
$$\phi(x) = -2^{2n+3}x + 2^{n+3} \quad \text{if } x \in]2^{-n-1}, 2^{-n}]$$
......

and we take $c = 1$ and

$$D = \{(x, y) \in \mathbb{R}^2 \ : \ x \in]0, 1[, \ y < \phi(x)\},$$

an optimal domain for the constrained isoperimetric problem does not exist. To see this fact it is enough to consider the minimizing sequence

$$A_n = \{(x, y) \in \mathbb{R}^2 \ : \ x \in]2^{-n-1}, 2^{-n}[, \ y < \phi(x)\}.$$

All the sets A_n are admissible and their Lebesgue measure is equal to 1 for all n; however, we have $\text{Per}_D(A_n) = 2^{-n-1} \to 0$, so that the infimum of the problem is zero. No optimal domain may then exist, because for every admissible set A we have $\text{Per}_D(A) > 0$. A picture of the set D is in the figure below.

The results above still hold for more general cost functionals. Instead of considering the cost given by the perimeter $|\nabla 1_A|$, take a function $j : \mathbb{R}^N \times \mathbb{R}^N \to \overline{\mathbb{R}}$ which satisfies the following properties:

i) j is lower semicontinuous on $\mathbb{R}^N \times \mathbb{R}^N$;

ii) for every $x \in \mathbb{R}^N$ the function $j(x, \cdot)$ is convex;

iii) there exists a constant $c_0 > 0$ such that

$$j(x, z) \geq c_0|z| \quad \forall (x, z) \in \mathbb{R}^N \times \mathbb{R}^N.$$

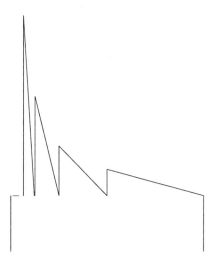

Figure 1.2: The constraint D.

Consider now the cost functional

$$F(A) = \int j(x, \nabla 1_A) \, .$$

The integral above must be intended in the sense of functionals over measures; more precisely, if μ is a measure and if $\mu = \mu^a \, dx + \mu^s$ is the Lebesgue-Nikodym decomposition of μ into absolutely continuous and singular parts (with respect to the Lebesgue measure), the integral $\int j(x, \mu)$ stands for

$$\int j(x, \mu^a(x)) \, dx + \int j^\infty \left(x, \frac{d\mu^s}{d|\mu^s|}\right) d\mu^s$$

where $|\mu^s|$ is the total variation of μ^s, $\frac{d\mu^s}{d|\mu^s|}$ is the Radon-Nikodym derivative of μ^s with respect to $|\mu^s|$, and j^∞ is the recession function of j defined by

$$j^\infty(x, z) = \lim_{t \to +\infty} \frac{j(x, tz)}{t} \, .$$

When $\mu = \nabla 1_A$ the expression above may be simplified; indeed, if A is a smooth domain it is easy to see that

$$\nabla 1_A = -\nu(x) \mathcal{H}^{N-1} \llcorner \partial A$$

being ν the exterior unit normal vector to A and $\mathcal{H}^{N-1}$ the $N-1$ dimensional Hausdorff measure. When A is not smooth, the correct way to represent the measure $\nabla 1_A$ is to introduce the so called *reduced boundary* $\partial^* A$.

Definition 1.2.7 *Let A be a set of finite perimeter; we say that $x \in \partial^* A$ if*

  i) for every $r > 0$ we have $0 < \text{meas} \left(A \cap B_{x,r} \right) < \text{meas} \left(B_{x,r} \right)$;

  ii) there exists the limit

$$\nu_A(x) = \lim_{r \to 0} \frac{-\nabla 1_A \left(B_{x,r} \right)}{|\nabla 1_A| \left(B_{x,r} \right)}$$

  and $|\nu_A(x)| = 1$.

The vector $\nu_A(x)$ is called exterior unit normal vector to A and the set $\partial^* A$ is called reduced boundary of A.

In this way, for every set A of finite perimeter we still have

$$\nabla 1_A = -\nu_A(x)\mathcal{H}^{N-1} \llcorner \partial^* A,$$

so that the cost functional above, dropping the constant term $\int_D j(x, 0)\, dx$, can be written as

$$F(A) = \int_{\partial^* A} j^\infty(x, -\nu_A)\, d\mathcal{H}^{N-1}.$$

It has to be noticed that the integrand $j^\infty(x, z)$ is positively homogeneous of degree 1 with respect to z. In an analogous way we may consider the functional

$$F_D(A) = \int_{D \cap \partial^* A} j^\infty(x, -\nu_A)\, d\mathcal{H}^{N-1}.$$

Theorem 1.2.8 *With the notation above, if the classes of admissible sets are nonempty, then the minimization problems*

$$\min \left\{ F(A) \;:\; A \subset K, \; \int_A f(x)\, dx = c \right\}$$
$$\min \left\{ F_D(A) \;:\; A \subset D, \; \int_A f(x)\, dx = c \right\}$$

both admit at least a solution, provided K is a compact set and D is a bounded open set.

Proof The proof in this more general framework is similar to the previous ones of Theorem 1.2.2 and Theorem 1.2.5. In fact, thanks to assumption iii) we still have the coercivity in the space BV, and thanks to assumptions i) and ii) the functionals F and F_D are lower semicontinuous with respect to the weak* convergence on BV (see for instance [59], [54]). ∎

We can now see how the boundary variation method works in the isoperimetric problem and how this allows us to obtain necessary conditions of optimality. Assume A is a solution of the isoperimetric problem

$$(1.5) \qquad \min \left\{ \operatorname{Per}_D(A) \; : \; A \subset D, \; \operatorname{meas}(A) = c \right\}.$$

and let $x_0 \in D \cap \partial A$; we assume that near x_0 the boundary ∂A is regular enough to perform all necessary operations. Actually, the regularity of ∂A does not need to be assumed as a hypothesis but is a consequence of some suitable conditions on the datum f; this is a quite delicate matter which goes under the name of *regularity theory*. We do not enter this field and we refer the interested reader to the various books available in the literature (see references [155], [126], [8]).

We can then assume that in a small neighbourhood of x_0 the boundary ∂A can be written as the graph of a function $u(x)$, where x varies in an open subset D of $\mathbb{R}^{N-1}$. The corresponding part of $\operatorname{Per}_D(A)$ can then be written in the Cartesian form as

$$\int_\omega \sqrt{1 + |\nabla u|^2} \, dx.$$

The boundary variation method consists in perturbing ∂A, hence $u(x)$, by taking a comparison function of the form $u(x) + \varepsilon\phi(x)$, where $\varepsilon > 0$ and ϕ is a smooth function with support in ω. We also want that the measure constraint remains fulfilled, which turns out to require that the function ϕ satisfies the equality

$$\int_\omega \phi(x) \, dx = 0.$$

Since A is optimal we obtain the inequality

$$(1.6) \qquad \int_\omega \sqrt{1 + |\nabla u + \varepsilon\nabla\phi|^2} \, dx \geq \int_\omega \sqrt{1 + |\nabla u|^2} \, dx.$$

The integrand in the left-hand side of (1.6) gives, as $\varepsilon \to 0$,

$$\sqrt{1 + |\nabla u + \varepsilon \nabla \phi|^2} = \sqrt{1 + |\nabla u|^2} + \varepsilon \frac{\nabla u \cdot \nabla \phi}{\sqrt{1 + |\nabla u|^2}} + o(\varepsilon^2)$$

so that (1.6) becomes

$$\int_\omega \frac{\nabla u \cdot \nabla \phi}{\sqrt{1 + |\nabla u|^2}}\, dx \geq 0.$$

Integrating by parts we obtain

$$- \int_\omega \operatorname{div} \left(\frac{\nabla u}{\sqrt{1 + |\nabla u|^2}} \right) \phi\, dx \geq 0,$$

and recalling that ϕ was arbitrary and with zero average in ω, we finally obtain that the function u must satisfy the partial differential equation

$$(1.7) \qquad - \operatorname{div} \left(\frac{\nabla u}{\sqrt{1 + |\nabla u|^2}} \right) = \text{constant} \qquad \text{in } \omega.$$

The term $- \operatorname{div} \left(\nabla u / \sqrt{1 + |\nabla u|^2} \right)$ represents the mean curvature of ∂A written in Cartesian coordinates; therefore we have found the following necessary condition of optimality for a regular solution A of the isoperimetric problem (1.5):

the mean curvature of $\Omega \cap \partial A$ is locally constant.

A more careful inspection of the proof above actually shows that the constant is the same on all $D \cap \partial A$. Indeed, if x_1 and x_2 are two points with neighbourhoods ω_1 and ω_2, and $u(x)$ is a function whose graph is ∂A in $\omega_1 \cup \omega_2$, the computation above gives

$$- \operatorname{div} \left(\frac{\nabla u}{\sqrt{1 + |\nabla u|^2}} \right) = c_1 \quad \text{in } \omega_1, \qquad - \operatorname{div} \left(\frac{\nabla u}{\sqrt{1 + |\nabla u|^2}} \right) = c_2 \quad \text{in } \omega_2$$

with c_1 and c_2 constants. Take now as a perturbation the function $u + \varepsilon(\phi_1 + \phi_2)$ where ϕ_1, ϕ_2 are smooth and with support in ω_1, ω_2 respectively. The measure constraint gives

$$(1.8) \qquad \int_{\omega_1} \phi_1\, dx + \int_{\omega_2} \phi_2\, dx = 0.$$

By repeating the argument used above we obtain

$$0 \leq \int_{\omega_1} \frac{\nabla u \cdot \nabla \phi_1}{\sqrt{1 + |\nabla u|^2}} \, dx + \int_{\omega_2} \frac{\nabla u \cdot \nabla \phi_2}{\sqrt{1 + |\nabla u|^2}} \, dx$$
$$= c_1 \int_{\omega_1} \phi_1 \, dx + c_2 \int_{\omega_2} \phi_2 \, dx.$$

Since ϕ_1 and ϕ_2 are arbitrary, with the only constraint (1.8), we easily obtain that $c_1 = c_2$, and so the mean curvature of $D \cap \partial A$ is globally a constant.

When the measure constraint is replaced by the more general constraint $\int_A f(x) \, dx = c$ we may easily repeat all the previous steps and we obtain the partial differential equation

$$(1.9) \qquad - \operatorname{div} \left(\frac{\nabla u}{\sqrt{1 + |\nabla u|^2}} \right) = \lambda f(x, u(x))$$

where λ is a constant.

Finally, when the perimeter is replaced by the more general functional

$$\int_{D \cap \partial^* A} j^\infty(x, \nu_A) \, d\mathcal{H}^{N-1},$$

then the exterior unit normal vector ν_A is, when ∂A is the graph of a smooth function u,

$$\nu_A = \left(-\frac{\nabla u}{\sqrt{1 + |\nabla u|^2}}, \frac{1}{\sqrt{1 + |\nabla u|^2}} \right)$$

so that the cost functional takes the form

$$\int_\omega j^\infty(x, u(x), \nabla u(x), 1) \, dx.$$

In this case the partial differential operator $-\operatorname{div} \left(\nabla u / \sqrt{1 + |\nabla u|^2} \right)$ has to be replaced by the new one obtained through the function $j^\infty(x, s, z, 1)$ that is

$$- \operatorname{div} \left(\partial_z j^\infty(x, u, \nabla u, 1) \right) + \partial_s j^\infty(x, u, \nabla u, 1).$$

1.3   The Newton problem of minimal aerodynamical resistance

The problem of finding the shape of a body which moves in a fluid with minimal resistance to motion is one of the first problems in the calculus of variations (see for instance Goldstine [130]). This can be again seen as a shape optimization problem, once the cost functional and the class of admissible shapes are defined.

In 1685 Newton studied this problem proposing a very simple model to describe the resistance of a profile to the motion in an inviscid and incompressible medium. Here are his words (from *Principia Mathematica*):

> *If in a rare medium, consisting of equal particles freely disposed at equal distances from each other, a globe and a cylinder described on equal diameter move with equal velocities in the direction of the axis of the cylinder, (then) the resistance of the globe will be half as great as that of the cylinder. ... I reckon that this proposition will be not without application in the building of ships.*

The Newtonian pressure law states that the pressure coefficient is proportional to $\sin^2 \theta$, being θ the inclination of the body profile with respect to the stream direction. The deduction of this pressure law can be easily obtained from the assumption that the fluid consists of many independent particles with constant speed and velocity parallel to the stream direction, the interactions between the body and the particles obey to the usual laws governing elastic shocks, and tangential friction and other effects are neglected (see figure below).

Suppose that the profile of the body is described by the graph of a nonnegative function u defined over the body cross section D (orthogonal to the fluid stream). A simple calculation gives that the effect due to the impact of a single particle, which slows the body down, that is the momentum in vertical direction, is proportional to the mass of the particle times $\sin^2 \theta$. Since

$$\sin^2 \theta = \frac{1}{1 + \tan^2(\pi/2 - \theta)} = \frac{1}{1 + |\nabla u|^2} \,,$$

the total resistance of the body turns out to be proportional to the integral

(1.10)
$$F(u) = \int_D \frac{1}{1 + |\nabla u|^2} \, dx.$$

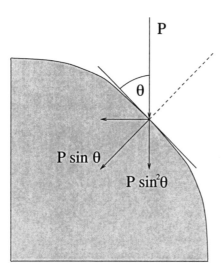

Figure 1.3: The Newtonian pressure law.

We may also define the relative resistance of a profile u, dividing the resistance $F(u)$ by the measure of D:

$$C_0(u) = \frac{F(u)}{|D|} .$$

In particular, if the body is a half-sphere of radius R we have $u(x) = \sqrt{R^2 - |x|^2}$ and an easy calculation gives the relative resistance

$$C_0(u) = \frac{F(u)}{\pi R^2} = 0.5$$

as predicted by Newton in 1685. Other bodies with the same value of C_0 are illustrated in figures below.

If we assume the total resistance to be our cost functional, it remains to determine the class of admissible shapes, that is the class of admissible functions u, over which the functional F has to be minimized.

Note that the integral functional F above is neither convex nor coercive. Therefore, obtaining an existence theorem for minimizers via the usual direct methods in the calculus of variations may fail.

If we do not impose any further constraint on the competing functions u, the infimum of the functional in (1.10) turns out to be zero, as it is immediate

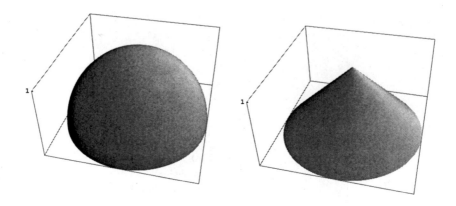

Figure 1.4: (a) half-sphere, (b) cone.

to see by taking for instance

$$u_n(x) = n \, \text{dist}(x, \partial D)$$

for every $n \in \mathbb{N}$ and by letting $n \to +\infty$. Therefore, no function u can minimize the functional F, because $F(u) > 0$ for every function u.

One may think that the nonexistence of minimizers for F is due to the fact that the sequence $\{u_n\}$ above is unbounded in the L^∞ norm; however, even a constraint of the form

(1.11) $0 \leq u \leq M$

does not help a lot for the existence of minimizers. Indeed, a sequence of functions like

$$u_n(x) = M \sin^2(n|x|)$$

satisfies the constraint (1.11) but we still have

$$\lim_{n \to +\infty} F(u_n) = 0,$$

and by the same argument used before we may conclude that again the cost functional F does not possess any minimizer, even in the more restricted class (1.11).

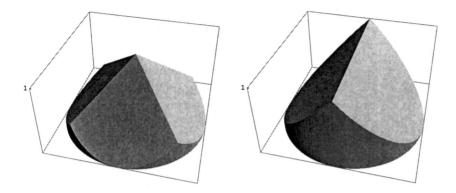

Figure 1.5: (c) pyramid 1, (d) pyramid 2.

We shall take as admissible bodies only convex bounded domains, that is we restrict our analysis to functions u which are bounded and concave on D. More precisely, we study the minimization problem

(1.12) $$\min\left\{F(u) \ : \ 0 \leq u \leq M, \ u \text{ concave on } D\right\}.$$

We shall see in Chapter 2 that the concavity constraint on u is strong enough to provide an extra compactness which implies the existence of a minimizer. On the other hand, from the physical point of view, a motivation for this constraint is that, thinking of the fluid as composed by many independent particles, each particle hits the body only once. If the body is not convex, it could happen that a particle hits the body more that once, but since $F(u)$ was constructed to measure only the resistance due to the first shock, it would no longer reflect the total resistance of the body.

Other kinds of constraints different from (1.11) can be imposed on the class of nonnegative concave functions: for instance, instead of (1.11) we may consider a bound on the surface area of the body, like

$$\int_D \sqrt{1 + |\nabla u|^2}\, dx + \int_{\partial D} u\, dH^{n-1} \leq c,$$

or on its volume, like

$$\int_D u\, dx \leq c.$$

We refer for instance to Miele's book [157] for a source of applications in Aerodynamics, and to some more recent papers ([137], [22], [181]).

Other classes of functions u, even if less motivated physically, can be considered from the mathematical point of view. A possibility could be the class of quasiconcave functions, that is of functions u whose upper level sets $\{x \in D : u(x) \geq t\}$ are all convex. Note that in the radially symmetric case a function $u = u(|x|)$ is quasiconcave if and only if it is decreasing as a function of $|x|$. Another class of admissible functions for which the problem can be studied is the class of superharmonic functions. Also the class of functions u with the property that the incoming particles hit the body only once deserves some interest. It is not the purpose of these notes to develop all details of these cases; thus we limit ourselves to the case of convex bodies, and we refer the interested reader to several papers where different situations are considered (see for instance [85], [86], [70], [66], [147], [76], [148]).

The most studied case of the Newton problem of profile with minimal resistance is when the competing functions are supposed a priori with a radial symmetry, that is D is a (two-dimensional) disk and the functions u only depend on the radial variable $|x|$. In this case, after integration in polar coordinates, the functional F can be written in the form

$$F(u) = 2\pi \int_0^R \frac{r}{1 + |u'(r)|^2}\, dr$$

so that the resistance minimization problem becomes

$$(1.13) \qquad \min\left\{ \int_0^R \frac{r}{1 + |u'(r)|^2}\, dr \ : \ u \text{ concave, } 0 \leq u \leq M \right\}.$$

Several facts about the radial Newton problem can be shown; we simply list them by referring to [70], [66], [68] for all details.

- It is possible to show that the competing functions $u(r)$ must satisfy the conditions $u(0) = M$ and $u(R) = 0$; moreover the infimum does not change if we minimize over the larger class of decreasing functions. Therefore problem (1.13) can also be written as

$$(1.14) \qquad \min\left\{ \int_0^R \frac{r}{1 + |u'(r)|^2}\, dr \ : \right.$$
$$\left. u \text{ decreasing, } u(0) = M, \ u(R) = 0 \right\}.$$

Notice that, when the function u is not absolutely continuous, the symbol u' under the integral in (1.14) stands for the absolutely continuous part of u'.

- By using the functions $v(t) = u^{-1}(M-t)$, problem (1.14) can be rewritten in the more traditional form

(1.15) $\quad \min \left\{ \int_0^M \frac{vv'^3}{1+v'^2} \, dr \; : \; v \text{ increasing}, \; v(0) = 0, \; v(M) = R \right\}.$

Again, when v is a general increasing function, v' is a nonnegative measure, and (1.14) has to be intended in the sense of BV functions, as

(1.16) $\qquad\qquad \int_0^M \frac{vv_a'^3}{1+v_a'^2} \, dt + \int_{[0,M]} vv_s'$

where v_a' and v_s' are respectively the absolutely continuous and singular parts of the measure v' with respect to Lebesgue measure. The second integral in (1.16) has the product vv_s' which may have some ambiguity in its definition: it is then better to add and subtract vv_a' so that the functional in (1.16) can be written in a simpler way as

$$\frac{R^2}{2} - \int_0^M \frac{vv_a'}{1+v_a'^2} \, dt.$$

- The minimization problem (1.14) admits an Euler-Lagrange equation which is, in its integrated form,

(1.17) $\qquad\qquad ru' = C\left(1+u'^2\right)^2 \qquad \text{on } \{u' \neq 0\}$

for a suitable constant $C < 0$. From (1.17) the solution u can actually be explicitly computed. Indeed, consider the function

$$f(t) = \frac{t}{(1+t^2)^2} \left(-\frac{7}{4} + \frac{3}{4}t^4 + t^2 - \log t \right) \qquad \forall t \geq 1;$$

we can easily verify that f is strictly increasing so that the following quantities are well defined:

$$T = f^{-1}(M/R)$$

$$r_0 = \frac{4RT}{(1+T^2)^2}.$$

Then we obtain

$$u(r) = M \qquad \forall r \in [0, r_0]$$

and the solution u can be computed in the parametric form

$$\begin{cases} r(t) = \dfrac{r_0}{4t}(1 + t^2)^2 \\ u(t) = M - \dfrac{r_0}{4}\left(-\dfrac{7}{4} + \dfrac{3}{4}t^4 + t^2 - \log t \right) \end{cases} \forall t \in [1, T].$$

Notice that $|u'(r)| > 1$ for all $r > r_0$ and that $|u'(r_0^+)| = 1$; in particular, the derivative $|u'|$ never belongs to the interval $]0, 1[$.

- The optimal radial shape for $M = R$ is shown in figure below.

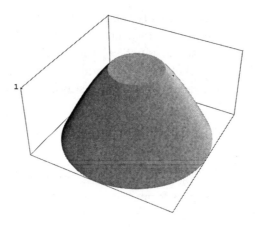

Figure 1.6: The optimal radial shape for $M = R$.

- It is possible to show that the optimal radial solution is unique.

- The optimal relative resistance C_0 of a radial body is then given by

$$C_0 = \frac{2}{R^2} \int_0^R \frac{r}{1 + u'^2} \, dr$$

where u is the optimal solution above. We have $C_0 \in [0, 1]$ and it is easy to see that C_0 depends on M/R only. Some approximate calculations give

	$M/R=1$	$M/R=2$	$M/R=3$	$M/R=4$
r_0/R	0.35	0.12	0.048	0.023
C_0	0.37	0.16	0.082	0.049

- Moreover we obtain the following asymptotic estimates as $M/R \to +\infty$:

(1.18)
$$r_0/R \approx \tfrac{27}{16}(M/R)^{-3} \qquad \text{as } M/R \to +\infty$$
$$C_0 \approx \tfrac{27}{32}(M/R)^{-2} \qquad \text{as } M/R \to +\infty.$$

Some more optimal radial shapes for different values of the ratio M/R are shown in the figure below.

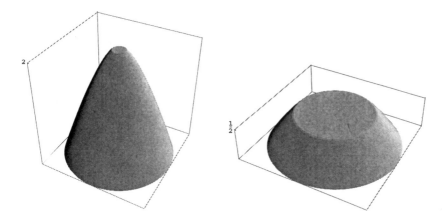

Figure 1.7: (a) the case $M = 2R$, (b) the case $M = R/2$.

- It is interesting to notice that the optimal frustum cone, that is the frustum cone with height M, cross section radius R, and minimal resistance, is only slightly less performant than the optimal radial body computed above. Indeed, its top radius $\hat{r}_0$ and its relative resistance $\hat{C}_0$ can be easily computed, and we find:

$$\hat{C}_0 = \frac{\hat{r}_0}{R} = 1 - \frac{(M/R)^2}{2}\left(\sqrt{1 + 4(M/R)^{-2}} - 1\right),$$

with asymptotic behaviour

$$\hat{C}_0 \approx (M/R)^{-2} \qquad \text{as } M/R \to +\infty.$$

In the nonradial case, we shall see in the next chapter that it is still possible to show the existence of an optimal profile, even if little is known about its qualitative behaviour. We shall see that a necessary condition of optimality is that the optimal profile must be flat, in the sense that $\det D^2 u$ identically vanishes where u is of class C^2. In particular, when D is a disk, this excludes the radial Newton solution and so the optimal solution cannot be radial. This also shows that the solution is not unique in general. Up to now it is not known if optimal solutions always have a *flat nose* and if they always assume the value zero at the boundary.

1.4   Optimal interfaces between two media

In this section we study the problem of finding the minimal energy configuration for a mixture of two conducting materials when a constraint (or penalization) on the measure of the unknown interface between the two phases is added.

If D denotes a given bounded open subset of $\mathbb{R}^N$ (the prescribed container), denoting by α and β the conductivities of the two materials, the problem consists in filling D with the two materials in the most performant way according to some given cost functional. The volume of each material can also be prescribed. It is convenient to denote by A the domain where the conductivity is α and by $a_A(x)$ the conductivity coefficient

$$a_A(x) = \alpha 1_A(x) + \beta 1_{D \setminus A}(x).$$

In this way the state equation becomes

(1.19)
$$\begin{cases} -\operatorname{div}\left(a_A(x)\nabla u\right) = f & \text{in } D \\ u = 0 & \text{on } \partial D, \end{cases}$$

where f is the (given) source density, and we denote by u_A its unique solution.

It is well known (see for instance Kohn and Strang [143], Murat and Tartar [164]) that if we take as a cost functional an integral of the form

$$\int_D j(x, 1_A, u_A, \nabla u_A)\, dx$$

in general an optimal configuration does not exist. However, the addition of

a perimeter penalization is enough to imply the existence of classical optimizers. In other words, if we take as a cost the functional

$$J(u, A) = \int_D j(x, 1_A, u, \nabla u) \, dx + \sigma \operatorname{Per}_D(A)$$

where $\sigma > 0$, the problem can be written as an optimal control problem in
the form

(1.20) $$\min \{ J(u, A) \ : \ A \subset D, \ u \text{ solves } (1.19) \}.$$

A volume constraint of the form meas$(A) = m$ could also be present. The
main ingredient for the proof of the existence of an optimal classical solution
is the following result.

Theorem 1.4.1 *Let $a_n(x)$ be a sequence of $N \times N$ symmetric matrices with
measurable coefficients such that the uniform ellipticity condition*

(1.21) $$c_0|z|^2 \leq a_n(x)z \cdot z \leq c_1|z|^2 \qquad \forall x \in D, \ \forall z \in \mathbb{R}^N$$

*holds with $0 < c_0 \leq c_1$. Given $f \in H^{-1}(D)$ denote by u_n the unique solution
of the problem*

(1.22) $$\begin{cases} -\operatorname{div} \left(a_n(x) \nabla u \right) = f \\ u \in H_0^1(D). \end{cases}$$

*If $a_n(x) \to a(x)$ a.e. in D then $u_n \to u$ weakly in $H_0^1(D)$, where u is the
solution of (1.22) with a_n replaced by a.*

Proof By the uniform ellipticity condition (1.21) we have

$$c_0 \int_D |\nabla u_n|^2 \, dx \leq \int_D f u_n \, dx$$

and, by Poincaré inequality we have that u_n are bounded in $H_0^1(D)$ so that a
subsequence (still denoted by the same indices) converges weakly in $H_0^1(D)$
to some v. All we have to show is that $v = u$ or equivalently that

(1.23) $$-\operatorname{div} \left(a(x) \nabla v \right) = f.$$

This means that for every smooth test function ϕ we have

$$\int_D a(x)\nabla v\nabla\phi\,dx = \langle f,\phi\rangle.$$

Then it is enough to show that for every smooth test function ϕ we have

$$\lim_{n\to+\infty}\int_D a_n(x)\nabla u_n\nabla\phi\,dx = \int_D a(x)\nabla v\nabla\phi\,dx.$$

This is an immediate consequence of the fact that ϕ is smooth, $\nabla u_n \to \nabla v$ weakly in $L^2(D)$, and $a_n \to a$ a.e. in D remaining bounded.

Another way to show that (1.23) holds is to verify that v minimizes the functional

$$(1.24)\qquad F(w) = \int_D a(x)\nabla w\nabla w\,dx - 2\langle f,w\rangle \qquad w \in H_0^1(D).$$

Since the function $\alpha(s,z) = sz\cdot z$ defined for $z \in \mathbb{R}^N$ and for s symmetric positive definite $N \times N$ matrix is convex in z and lower semicontinuous in s, the functional

$$\Phi(a,\xi) = \int_D a(x)\xi\cdot\xi\,dx$$

is sequentially lower semicontinuous with respect to the strong L^1 convergence on a and the weak L^1 convergence on ξ (see for instance [54]). Therefore we have

$$F(v) = \Phi(a,\nabla v) - 2\langle f,v\rangle \le \liminf_{n\to+\infty}\Phi(a_n,\nabla u_n) - 2\langle f,u_n\rangle = \liminf_{n\to+\infty}F(u_n).$$

Since u_n minimizes the functional F_n defined as in (1.24) with a replaced by a_n, we also have for every $w \in H_0^1(D)$

$$F_n(u_n) \le F_n(w) = \int_D a_n(x)\nabla w\nabla w\,dx - 2\langle f,w\rangle$$

so that taking the limit as $n \to +\infty$ and using the convergence $a_n \to a$ we obtain

$$\liminf_{n\to+\infty}F_n(u_n) \le \int_D a(x)\nabla w\nabla w\,dx - 2\langle f,w\rangle = F(w).$$

Thus $F(v) \le F(w)$ which shows what required. ∎

Remark 1.4.2 The result above can be rephrased in terms of G-convergence by saying that for uniformly elliptic operators of the form $-\operatorname{div}\left(a(x)\nabla u\right)$ the G-convergence is weaker than the L^1-convergence of coefficients. Analogously, we can say that the functionals

$$G_n(w) = \int_D a_n(x)\nabla w\nabla w\,dx$$

Γ-converge to the functional G defined in the same way with a in the place of a_n.

Corollary 1.4.3 *If $A_n \to A$ in $L^1(D)$ then $u_{A_n} \to u_A$ weakly in $H_0^1(D)$.*

A more careful inspection of the proof of Theorem 1.4.1 shows that the following stronger result holds.

Theorem 1.4.4 *Under the same assumptions of Theorem 1.4.1 the convergence of u_n is actually strong in $H_0^1(D)$.*

Proof We have already seen that $u_n \to u$ weakly in $H_0^1(D)$, which gives $\nabla u_n \to \nabla u$ weakly in $L^2(D)$. Denoting by $c_n(x)$ and $c(x)$ the square root matrices of $a_n(x)$ and $a(x)$ respectively, we have that $c_n \to c$ a.e. in D remaining equi-bounded. Then $c_n(x)\nabla u_n$ converge to $c(x)\nabla u$ weakly in $L^2(D)$. Multiplying equation (1.22) by u_n and integrating by parts we obtain

$$\int_D a(x)\nabla u\nabla u\,dx = \langle f, u\rangle = \lim_{n\to+\infty}\langle f, u_n\rangle$$
$$= \lim_{n\to+\infty}\int_D a_n(x)\nabla u_n\nabla u_n\,dx.$$

This implies that

$$c_n(x)\nabla u_n \to c(x)\nabla u \qquad \text{strongly in } L^2(D).$$

Multiplying now by $\left(c_n(x)\right)^{-1}$ we finally obtain the strong convergence of ∇u_n to ∇u in $L^2(D)$. ∎

We are now in a position to obtain an existence result for the optimization problem (1.20). On the function j we only assume that it is nonnegative, Borel measurable, and such that $j(x, s, z, w)$ is lower semicontinuous in (s, z, w) for a.e. $x \in D$.

Theorem 1.4.5 *Under the assumption above the minimum problem (1.20) admits at least a solution.*

Proof Let (A_n) be a minimizing sequence; then $\mathrm{Per}_D(A_n)$ are bounded, so that, up to extracting subsequences, we may assume (A_n) is strongly convergent in the L^1_{loc} sense to some set $A \subset D$. We claim that A is a solution of problem (1.20). Let us denote by u_n a solution of problem (1.19) associated to A_n; by Theorem 1.4.4 (u_n) converges strongly in $H^1_0(D)$ to some $u \in H^1_0(D)$. Then by the lower semicontinuity of the perimeter and by Fatou's lemma we have

$$J(u, A) \leq \liminf_{n \to +\infty} J(u_n, A_n)$$

which proves the optimality of A.   ∎

Remark 1.4.6 The same proof works when volume constraints of the form $\mathrm{meas}(A) = m$ are present. Indeed this constraint passes to the limit when $A_n \to A$ strongly in $L^1(D)$.

The existence result above shows the existence of a classical solution for the optimization problem (1.20). This solution is simply a set with finite perimeter and additional assumptions have to be made in order to prove further regularity. For instance in [11] Ambrosio and Buttazzo considered the similar problem

$$\min \left\{ J(u, A) \ : \ u \in H^1_0(D), \ A \subset D \right\}$$

where

$$J(u, A) = \int_D \left[a_A(x)|\nabla u|^2 + 1_A(x)g_1(x, u) + 1_{D \setminus A} g_2(x, u) \right] dx$$

and showed that every solution A is actually an open set provided g_1 and g_2 are Borel measurable and satisfy the inequalities

$$g_i(x, s) \geq \gamma(x) - k|s|^2 \qquad i = 1, 2$$

where $\gamma \in L^1(D)$ and $k < \alpha \lambda_1$, being λ_1 the first eigenvalue of $-\Delta$ on D.

1.5 The optimal shape of a thin insulating layer

In this section we study the optimization problem for a thin insulating layer around a conducting body; we have to put a given amount of insulating material on the boundary of a given domain in order to minimize a cost functional which describes the total heat dispersion of the domain. We consider the framework of stationary heat equation but the same model also applies to similar problems in electrostatics or in the case of elastic membranes.

Let D be a regular bounded open subset of $\mathbb{R}^N$ that we suppose connected for simplicity and let $f \in L^2(D)$ be a given function which represents the heat sources density. We assume that the boundary ∂D is surrounded by a thin layer of insulator, with thickness $d(\sigma)$ being σ the variable which runs over ∂D. The limit problem, when the thickness of the layer goes to zero and simultaneously its insulating coefficient goes to infinity (i.e. the conductivity in the layer goes to zero too), has been studied in [31] through a PDE approach (and called *reinforcement problem*) and in [1] through a Γ-limit approach, and the model obtained is the following. If u denotes the temperature of the system, then the family of approximating problems is

$$(1.25) \quad \min\left\{ \int_D |\nabla u|^2 \, dx + \varepsilon \int_{\Sigma_\varepsilon} |\nabla u|^2 \, dx - 2 \int_D fu \, dx \ : \ u \in H_0^1(D \cup \Sigma_\varepsilon) \right\}$$

where Σ_ε is the thin layer of variable thickness $d(\sigma)$

$$(1.26) \qquad \Sigma_\varepsilon = \left\{ \sigma + t\nu(\sigma) \ : \ \sigma \in \partial D, \ 0 \leq t < \varepsilon d(\sigma) \right\}.$$

In terms of PDE the Euler-Lagrange equation associated to problem (1.25) is an elliptic problem with a transmission condition along the interface ∂D

$$\begin{cases} -\Delta u = f & \text{in } D \\ -\Delta u = 0 & \text{in } \Sigma_\varepsilon \\ \frac{\partial u^-}{\partial \nu} = \varepsilon \frac{\partial u^+}{\partial \nu} & \text{on } \partial D \\ u = 0 & \text{on } \partial(D \cup \Sigma_\varepsilon) \end{cases}$$

where u^- and u^+ respectively denote the traces of u in D and in Σ_ε.

Notice that the conductivity coefficient in the insulating layer Σ_ε has been taken equal to ε, as well as the size of the layer thickness. Passing to the limit as $\varepsilon \to 0$ (in the sense of Γ-convergence) in the sequences of energy functionals we obtain (see [1]) the limit energy which is given by

$$(1.27) \qquad E(u, d) = \int_D |\nabla u|^2 \, dx - 2 \int_D fu \, dx + \int_{\partial D} \frac{u^2}{d} \, d\mathcal{H}^{N-1}$$

so that the temperature u solves the minimum problem

$$(1.28) \qquad E(d) = \min \left\{ E(u, d) \ : \ u \in H^1(D) \right\}.$$

Equivalently, problem (1.28) can be described through its Euler-Lagrange equation

$$(1.29) \qquad \begin{cases} -\Delta u = f & \text{in } D \\ d \frac{\partial u}{\partial \nu} + u = 0 & \text{on } \partial D. \end{cases}$$

We denote by u_d the unique solution of (1.28) or of (1.29). Equation (1.29) can be seen as the state equation of an optimal control problem whose state variable is the temperature of the system and whose control variable is the thickness function $d(\sigma)$. Given a fixed amount m of insulating material the control variables we consider are (measurable) thickness functions d defined on ∂D such that

$$d \geq 0 \text{ on } \partial D, \qquad \int_{\partial D} d \, d\mathcal{H}^{N-1} = m.$$

We denote by Γ_m such a class of functions. Therefore, the optimization problem we are going to consider is

$$(1.30) \quad \min \left\{ E(d) \ : \ d \in \Gamma_m \right\} = \min \left\{ E(u, d) \ : \ u \in H^1(D), \ d \in \Gamma_m \right\}.$$

Remark 1.5.1 The energy $E(d)$ in (1.28) can be written in terms of the solution u_d; indeed, multiplying equation (1.29) by u_d and integrating by parts, we obtain

$$(1.31) \qquad E(d) = E(u_d, d) = - \int_D fu_g \, dx.$$

Therefore, when the heat sources are uniformly distributed, that is f is (a positive) constant, the optimization problem (1.30) turns out to be equivalent to determining the function $d \in \Gamma_m$ for which the averaged temperature

$\int_D u_d \, dx$ is maximal. Other criteria, different from the minimization of the energy $E(d)$, could be also investigated, as for instance obtaining a temperature as close as possible to a desired state $a(x)$

$$\min\left\{ \int_D |u_d - a(x)|^2 \, dx \; : \; d \in \Gamma_m \right\}$$

or more generally

$$\min\left\{ \int_D f(x, u_d) \, dx + \int_{\partial D} g(x, d, u_d) \, d\mathcal{H}^{N-1} \; : \; d \in \Gamma_m \right\}.$$

For further details we refer to the chapters of these notes where we consider the general theory of shape optimization for problems with Dirichlet condition on the free boundary.

Proposition 1.5.2 *For every $u \in L^2(\partial D)$ the minimum problem*

(1.32)
$$\min\left\{ \int_{\partial D} \frac{u^2}{d} \, d\mathcal{H}^{N-1} \; : \; d \in \Gamma_m \right\}$$

admits a solution. This solution is unique if u is not identically zero.

Proof If $u = 0$, then any function $d \in \Gamma_m$ solves the minimization problem (1.32). Assume that u is nonzero; then we claim that the function

$$d_u = m|u| \left(\int_{\partial D} |u| \, d\mathcal{H}^{N-1} \right)^{-1}$$

solves the minimization problem (1.32). In fact, by Hölder inequality we have, for every $d \in \Gamma_m$

$$\left(\int_{\partial D} |u| \, d\mathcal{H}^{N-1} \right)^2 \le \left(\int_{\partial D} \frac{u^2}{d} \, d\mathcal{H}^{N-1} \right) \left(\int_{\partial D} d \, d\mathcal{H}^{N-1} \right) = m \int_{\partial D} \frac{u^2}{d} \, d\mathcal{H}^{N-1}$$

so that

$$\int_{\partial D} \frac{u^2}{d_u} \, d\mathcal{H}^{N-1} = \frac{1}{m} \left(\int_{\partial D} |u| \, d\mathcal{H}^{N-1} \right)^2 \le \int_{\partial D} \frac{u^2}{d} \, d\mathcal{H}^{N-1}$$

which proves the optimality of d_u. The uniqueness of the solution follows from the strict convexity of the mapping $d \mapsto 1/d$ and from the fact that every solution of (1.32) must vanish on the set $\{x \in \partial D \; : \; u(x) = 0\}$. ∎

Interchanging the order of the minimization in problem (1.30) we can perform first the minimization with respect to d, so that, thanks to the result of Proposition 1.5.2, problem (1.30) reduces to

(1.33)
$$\min\left\{\int_D |\nabla u|^2\,dx - 2\int_D fu\,dx + \frac{1}{m}\left(\int_{\partial D} |u|\,d\mathcal{H}^{N-1}\right)^2 : u \in H^1(D)\right\}.$$

It is immediate to see that the variational problem above is convex; then it can equivalently be seen in terms of its Euler-Lagrange equation which has the form

$$\begin{cases} -\Delta u = f & \text{in } D \\ 0 \in m\frac{\partial u}{\partial \nu} + H(u)\int_{\partial D}|u|\,d\mathcal{H}^{N-1} & \text{on } \partial D \end{cases}$$

where $H(t)$ denotes the multimapping

$$H(t) = \begin{cases} 1 & \text{if } t > 0 \\ -1 & \text{if } t < 0 \\ [-1,1] & \text{if } t = 0 \end{cases}$$

The following Poincaré-type inequality will be useful.

Proposition 1.5.3 *There exists a constant C such that for every $u \in H^1(D)$*

(1.34)
$$\int_D u^2\,dx \le C\left[\int_D |\nabla u|^2\,dx + \left(\int_{\partial D}|u|\,d\mathcal{H}^{N-1}\right)^2\right].$$

Proof If we assume by contradiction that (1.34) is false we may find a sequence (u_n) in $H^1(D)$ such that

$$\int_D u_n^2\,dx = 1, \qquad \int_D |\nabla u_n|^2\,dx + \left(\int_{\partial D}|u_n|\,d\mathcal{H}^{N-1}\right)^2 \to 0.$$

Possibly passing to subsequences we may then assume that u_n converge weakly in $H^1(D)$ to some $u \in H^1(D)$ with $\int_D u^2\,dx = 1$. Since $\int_D |\nabla u_n|^2\,dx \to 0$ the convergence is actually strong in $H^1(D)$ and since $u_n \to u$ strongly in $L^2(\partial D)$ we have that

$$\nabla u \equiv 0 \text{ in } D, \qquad u \in H^1_0(D).$$

The proof is then concluded because this implies $u \equiv 0$ which is in contradiction with the fact that $\int_D u^2\,dx = 1$. ∎

Proposition 1.5.4 *For every $f \in L^2(D)$ the minimization problem (1.33) admits a unique solution.*

Proof Let (u_n) be a minimizing sequence of problem (1.33); by comparison with the null function we have

$$\int_D |\nabla u_n|^2 \, dx - 2 \int_D f u_n \, dx + \frac{1}{m} \left(\int_{\partial D} |u_n| \, d\mathcal{H}^{N-1} \right)^2 \le 0.$$

Therefore, by using Hölder inequality, for every $\varepsilon > 0$ we have

$$\int_D |\nabla u_n|^2 \, dx + \frac{1}{m} \left(\int_{\partial D} |u_n| \, d\mathcal{H}^{N-1} \right)^2 \le 2 \int_D |f u_n| \, dx$$
$$\le 2 \|f\|_{L^2(D)} \|u_n\|_{L^2(D)} \le \frac{1}{\varepsilon} \int_D |f|^2 \, dx + \varepsilon \int_D |u_n|^2 \, dx$$

for every $n \in \mathbb{N}$. By using the Poincaré-type inequality of Proposition 1.5.3 we obtain

$$\int_D |\nabla u_n|^2 \, dx + \frac{1}{m} \left(\int_{\partial D} |u_n| \, d\mathcal{H}^{N-1} \right)^2 \le \frac{1}{\varepsilon} \int_D |f|^2 \, dx$$
$$+ \varepsilon C \left[\int_D |\nabla u_n|^2 \, dx + \frac{1}{m} \left(\int_{\partial D} |u_n| \, d\mathcal{H}^{N-1} \right)^2 \right]$$

so that, by taking ε sufficiently small, (u_n) turns out to be bounded in $H^1(D)$. Possibly passing to subsequences, we may assume $u_n \to u$ weakly in $H^1(D)$ for some function $u \in H^1(D)$, and the weak $H^1(D)$-lower semicontinuity of the energy functional

$$G(u) = \int_D |\nabla u|^2 \, dx - 2 \int_D f u \, dx + \frac{1}{m} \left(\int_{\partial D} |u| \, d\mathcal{H}^{N-1} \right)^2$$

gives that u is a solution of problem (1.33).

In order to prove the uniqueness, assume u_1 and u_2 are two different solutions of problem (1.33); a simple computation shows that

$$G\left(\frac{u_1 + u_2}{2} \right) - \frac{G(u_1) + G(u_2)}{2} = -\frac{1}{4} \int_D |\nabla u_1 - \nabla u_2|^2 \, dx$$
$$+ \frac{1}{4m} \left(\int_{\partial D} |u_1 + u_2| \, d\mathcal{H}^{N-1} \right)^2 - \frac{1}{2m} \left(\int_{\partial D} |u_1| \, d\mathcal{H}^{N-1} \right)^2$$
$$- \frac{1}{2m} \left(\int_{\partial D} |u_2| \, d\mathcal{H}^{N-1} \right)^2.$$

Moreover, the right-hand side is strictly negative whenever $u_1 - u_2$ is non-constant, which gives in this case a contradiction to the minimality of u_1 and u_2.

It remains to consider the case $u_1 - u_2 = c$ with c constant. If u_1 and u_2 have a different sign on a subset B of ∂D with $\mathcal{H}^{N-1}(B) > 0$, we have

$$|u_1 + u_2| < |u_1| + |u_2| \qquad \mathcal{H}^{N-1}\text{-a.e. on } B$$

which again contradicts the minimality of u_1 and u_2. If finally u_1 and u_2 have the same sign on ∂D we have

$$\left(\int_{\partial D} |u_1 + u_2| \, d\mathcal{H}^{N-1} \right)^2 - 2\left(\int_{\partial D} |u_1| \, d\mathcal{H}^{N-1} \right)^2 - 2\left(\int_{\partial D} |u_2| \, d\mathcal{H}^{N-1} \right)^2$$
$$= -\left(\int_{\partial D} |u_1 - u_2| \, d\mathcal{H}^{N-1} \right)^2 = -c^2 \mathcal{H}^{N-1}(\partial D)$$

which gives again a contradiction and concludes the proof. ∎

We are now in a position to prove an existence result for the optimization problem (1.30).

Theorem 1.5.5 *Let $f \in L^2(D)$ be fixed. Then the optimization problem (1.30) admits at least one solution d_{opt}. Moreover, denoting by u_{opt} the unique solution of (1.33), if u_{opt} does not vanish identically on ∂D we have that d_{opt} is unique and is given by*

$$d_{opt}(\sigma) = m|u_{opt}(\sigma)| \left(\int_{\partial D} |u_{opt}| \, d\mathcal{H}^{N-1} \right)^{-1} \qquad \text{for } \mathcal{H}^{N-1}\text{-a.e. } \sigma \in \partial D.$$

Proof The proof follows straightforward from Proposition 1.5.2 and Proposition 1.5.4. ∎

Remark 1.5.6 It is clear that, when u_{opt} identically vanishes on ∂D, any function $d \in \Gamma_m$ can be taken as a solution of the optimization problem (1.30). However, this does not occur, at least if f is a nonnegative (and not identically zero) function, as it is easy to see by comparing the energy of the Dirichlet solution u_0 to the energy of the function $u_0 + \varepsilon\phi$ with $\phi > 0$ and $\varepsilon > 0$ small enough. Moreover, problem (1.30) does not change if we replace the constraint $\int_{\partial D} d \, d\mathcal{H}^{N-1} = m$ by the constraint $\int_{\partial D} d \, d\mathcal{H}^{N-1} \leq m$. Finally, all the previous analysis still holds if the heat sources density f is taken in the dual space $\left(H^1(D) \right)'$.

Chapter 2

Optimization problems over classes of convex domains

In this chapter we deal with optimization problems whose class of admissible domains is made of convex sets. This geometrical constraint is rather strong and sufficient in many cases to guarantee the existence of an optimal solution.

In Section 2.1 the cost functional will be an integral functional of the form $\int_D f(x, u, \nabla u)\, dx$ where D is fixed and u varies in a class of convex (or concave, like in the case of Newton problem) functions on D. We shall see that the convexity conditions provides an extra compactness which gives the existence of an optimal domain under very mild conditions on the integrand f.

In Section 2.2 we consider the case of cost functionals which are boundary integrals of the form $\int_{\partial A} f(x, \nu)\, d\mathcal{H}^{N-1}$. Again, the convexity hypothesis on the admissible domains A will enable us to obtain the existence of an optimal solution.

Section 2.3 deals with some optimization problems governed by partial differential equations of higher order; the situations considered are such that the convexity condition is strong enough to provide the existence of a solution.

In all these cases it would be interesting to enlarge the class of convex domains by imposing some weaker geometrical conditions but still strong enough to give the existence of an optimal solution.

2.1   A general existence result for variational integrals

Starting from the discussion about the Newton problem of optimal aerodynamical profile made in Section 1.3, we consider in this section the general case of cost functionals of the form

$$F(u) = \int_D f(x, u, \nabla u)\, dx$$

where D is a given convex subset of $\mathbb{R}^N$ ($N = 2$ in the physical case) and the integrand f satisfies the very mild assumptions:

A1 the function $f : D \times \mathbb{R} \times \mathbb{R}^N \to \overline{\mathbb{R}}$ is nonnegative and measurable for the σ-algebra $\mathcal{L}_N \otimes \mathcal{B} \otimes \mathcal{B}_N$;

A2 for a.e. $x \in D$ the function $f(x, \cdot, \cdot)$ is lower semicontinuous on $\mathbb{R} \times \mathbb{R}^N$.

The case of Newton resistance functional is described by the integrand

$$f(z) = \frac{1}{1 + |z|^2} \ .$$

Note that no convexity assumptions on the dependence of $f(x, s, z)$ on z are made. This lack of convexity in the integrand does not allow us to apply the direct methods of the calculus of variations in its usual form, with a functional defined on a Sobolev space endowed with a weak topology (see [54], [88]).

The class of admissible functions u we shall work with is, like in the Newton problem, the class

$$C_M = \big\{u \text{ concave on } D \ : \ 0 \le u \le M\big\}$$

where $M > 0$ is a given constant. Other kind of classes are considered in the literature (see for instance [66], [85],[86], [147], [148], [21]).

The minimum problem we deal with is then

$$(2.1) \qquad\qquad \min\big\{F(u) \ : \ u \in C_M\big\}.$$

Note that, since every bounded concave function is locally Lipschitz continuous in D, the functional F in (2.1) is well defined on C_M. Moreover, as a consequence of Fatou's lemma, conditions **A1** and **A2** imply that the functional F is lower semicontinuous with respect to the strong convergence of every Sobolev space $W^{1,p}(D)$ or also $W^{1,p}_{loc}(D)$.

The result we want to prove is the following.

Theorem 2.1.1 *Under assumptions* **A1** *and* **A2**, *for every $M > 0$ the minimum problem*

$$(2.2) \qquad\qquad \min\left\{F(u) \ : \ u \in C_M\right\}$$

admits at least a solution.

The proof of the existence Theorem 2.1.1 relies on the following compactness result for the class C_M (see Marcellini [154]).

Lemma 2.1.2 *For every $M > 0$ and every $p < +\infty$ the class C_M is compact with respect to the strong topology of $W^{1,p}_{loc}(D)$.*

Proof Let (u_n) be a sequence of elements of C_M; since all u_n are concave, they are locally Lipschitz continuous on D, that is

$$\forall D' \subset\subset D \quad \forall x, y \in D' \quad |u_n(x) - u_n(y)| \leq C_{n,D'}|x - y|$$

where $C_{n,D'}$ is a suitable constant. Moreover, from the fact that $0 \leq u_n \leq M$, the constants $C_{n,D'}$ can be choosen independent of n; in fact it is well known that we can take

$$C_{n,D'} = 2M/\operatorname{dist}(D', \partial D).$$

Therefore the sequence (u_n) is equi-bounded and equi-Lipschitz continuous on every subset D' which is relatively compact in D. Thus, by Ascoli-Arzelà theorem, (u_n) is compact with respect to the uniform convergence in D' for every $D' \subset\subset D$. By a diagonal argument we may construct a subsequence of (u_n) (that we still denote by (u_n)) such that $u_n \to u$ uniformly on all compact subsets of D, for a suitable $u \in C_M$. Since the gradients ∇u_n are equi-bounded on every $D' \subset\subset D$, by the Lebesgue dominated convergence theorem, in order to conclude the proof it is enough to show that

$$(2.3) \qquad\qquad \nabla u_n(x) \to \nabla u(x) \qquad \text{for a.e. } x \in D.$$

Let us fix an integer $k \in [1, n]$ and a point $x \in D$ where all u_n and u are differentiable (since all u_n and u are locally Lipschitz continuous, almost all points $x \in D$ are of this kind). Now, the functions $t \mapsto u_n(x + te_k)$ are concave, so that we get for every $\varepsilon > 0$

$$(2.4) \qquad \frac{u_n(x + \varepsilon e_k) - u_n(x)}{\varepsilon} \leq \nabla_k u_n(x) \leq \frac{u_n(x - \varepsilon e_k) - u_n(x)}{-\varepsilon} \,,$$

where we denoted by e_k the k-th vector of the canonical orthogonal basis of $\mathbb{R}^N$. Passing to the limit as $n \to +\infty$ in (2.4) we obtain for every $\varepsilon > 0$

$$(2.5) \qquad \begin{aligned} \frac{u(x + \varepsilon e_k) - u(x)}{\varepsilon} &\leq \liminf_{n \to +\infty} \nabla_k u_n(x) \leq \\ &\leq \limsup_{n \to +\infty} \nabla_k u_n(x) \leq \frac{u(x - \varepsilon e_k) - u(x)}{-\varepsilon}. \end{aligned}$$

Passing now to the limit as $\varepsilon \to 0$ we finally have

$$\nabla_k u(x) \leq \liminf_{n \to +\infty} \nabla_k u_n(x) \leq \limsup_{n \to +\infty} \nabla_k u_n(x) \leq \nabla_k u(x)$$

that is (2.3), as required.   ∎

Proof [of Theorem 2.1.1] The existence result follows from the direct method of the calculus of variations. As we already noticed, thanks to assumptions **A1** and **A2** the functional F is lower semicontinuous with respect to the strong convergence of the Sobolev space $W^{1,p}_{loc}(D)$. By Lemma 2.1.2 the class C_M is also compact for the same convergence. This is enough to conclude that the minimum problem (2.2) admits at least a solution.   ∎

In particular, the problem of minimal Newtonian resistance

$$(2.6) \qquad \min \left\{ \int_D \frac{1}{1 + |\nabla u|^2} \, dx \ : \ u \in C_M \right\}$$

admits a solution for every $M \geq 0$.

A class larger than C_M that could be considered is the class of superharmonic functions

$$(2.7) \qquad E_M = \left\{ u \in H^1_{loc}(D) \ : \ 0 \leq u \leq M, \quad \Delta u \leq 0 \text{ in } D \right\}.$$

Here Δu is intended in the sense of distributions; then instead of requiring, as in the case C_M, that the $N \times N$ matrix $D^2 u$ is negative (as a measure), here we simply require that its trace Δu is negative. Nevertheless, we still have a compactness result as the following theorem shows.

Lemma 2.1.3 *Let (u_n) be a sequence of functions in E_M. Then for every $\alpha > 0$ there exists an open set $A_\alpha \subset D$ with $\mathrm{meas}(A_\alpha) < \alpha$ and a subsequence (u_{n_k}) such that ∇u_{n_k} converge strongly in $L^2_{loc}(D \setminus A_\alpha)$.*

Proof For every $\delta > 0$ let us denote by D_δ the set

$$D_\delta = \{x \in D \ : \ \mathrm{dist}(x, \partial D) > \delta\}.$$

Consider a smooth cut-off function η_δ with compact support in D and such that

$$0 \leq \eta_\delta \leq 1, \quad \eta_\delta = 1 \text{ on } D_\delta, \quad |\nabla \eta_\delta| \leq \frac{2}{\delta},$$

and set $\phi_{n,\delta} = \eta_\delta^2 (M - u_n)$. Since u_n are superharmonic we have

$$0 \leq \int_D \nabla u_n \nabla \phi_{n,\delta} \, dx = \int_D \left[2\eta_\delta (M - u_n)\nabla u_n \nabla \eta_\delta - \eta_\delta^2 |\nabla u_n|^2\right] dx$$

so that

$$(2.8) \quad \begin{aligned} \int_D \eta_\delta^2 |\nabla u_n|^2 \, dx &\leq \int_D 2\eta_\delta (M - u_n)|\nabla u_n||\nabla \eta_\delta| \, dx \\ &\leq \frac{1}{2} \int_D \eta_\delta^2 |\nabla u_n|^2 \, dx + 2\int_D (M - u_n)^2 |\nabla \eta_\delta|^2 \, dx. \end{aligned}$$

Hence

$$(2.9) \quad \begin{aligned} \int_{D_\delta} |\nabla u_n|^2 \, dx &\leq \int_D \eta_\delta^2 |\nabla u_n|^2 \, dx \\ &\leq 4\int_D (M - u_n)^2 |\nabla \eta_\delta|^2 \, dx \\ &\leq \frac{16 M^2 \, \mathrm{meas}(D)}{\delta^2} = C(\delta). \end{aligned}$$

Therefore (u_n) is bounded in $H^1(D_\delta)$ and so it has a subsequence weakly convergent to some $u \in E_M$ in $H^1(D_\delta)$. Possibly passing to subsequences, and by using a diagonal argument, we may assume that (u_n) converges strongly in $L^2(D)$. Using Egorov's theorem, for every $\alpha > 0$ there exists an open set $A_\alpha \subset D$ with $\mathrm{meas}(A_\alpha) < \alpha$ such that (u_n) converges uniformly on $D \setminus A_\alpha$. Fix now $\varepsilon > 0$ and define

$$v_n = (\varepsilon + u - u_n)^+;$$

since $\Delta u_n \leq 0$ we obtain

$$
\begin{aligned}
0 \;\leq\; & \int_D \nabla u_n \nabla(\eta_\delta^2 v_n)\, dx \\
(2.10) \qquad =\; & \int_{\{u_n - u \leq \varepsilon\}} \left[2\eta_\delta v_n \nabla u_n \nabla \eta_\delta + \eta_\delta^2 \nabla u_n \nabla v_n \right] dx \\
=\; & \int_{\{u_n - u \leq \varepsilon\}} \left[2\eta_\delta (\varepsilon + u - u_n) \nabla u_n \nabla \eta_\delta - \eta_\delta^2 \nabla u_n \nabla(u_n - u) \right] dx
\end{aligned}
$$

so that

$$
(2.11) \quad \int_{\{u_n - u \leq \varepsilon\}} \eta_\delta^2 \nabla u_n \nabla(u_n - u)\, dx \leq 2 \int_{\{u_n - u \leq \varepsilon\}} \eta_\delta(\varepsilon + u - u_n) \nabla u_n \nabla \eta_\delta.
$$

Since $u_n - u \leq \varepsilon$ on $D \setminus A_\alpha$, for n large enough, we have by (2.11)

$$
\begin{aligned}
\int_{D_\delta \setminus A_\alpha} |\nabla u_n - \nabla u|^2\, dx \;\leq\; & \int_{D \setminus A_\alpha} \eta_\delta^2 |\nabla u_n - \nabla u|^2\, dx \\
\leq\; & \int_{\{u_n - u \leq \varepsilon\}} \left[\eta_\delta^2 \nabla u_n \nabla(u_n - u) - \eta_\delta^2 \nabla u \nabla(u_n - u) \right] dx \\
(2.12) \qquad \leq\; & 2 \int_{\{u_n - u \leq \varepsilon\}} \eta_\delta(\varepsilon + u - u_n) |\nabla u_n| |\nabla \eta_\delta|\, dx \\
& - \int_{\{u_n - u \leq \varepsilon\}} \eta_\delta^2 \nabla u \nabla(u_n - u)\, dx.
\end{aligned}
$$

Since $\nabla u_n \to \nabla u$ weakly, the second integral in the last line tends to 0 as $n \to +\infty$, while

$$
\begin{aligned}
& \int_{\{u_n - u \leq \varepsilon\}} \eta_\delta(\varepsilon + u - u_n) |\nabla u_n| |\nabla \eta_\delta|\, dx \\
(2.13) \qquad & \leq \left(\int_D \eta_\delta^2 |\nabla u_n|^2\, dx \right)^{1/2} \left(\int_D |\varepsilon + u - u_n|^2 |\nabla \eta_\delta|^2\, dx \right)^{1/2} \\
& \leq \frac{2C(\delta)^{1/2}}{\delta} \left(\int_D |\varepsilon + u - u_n|^2\, dx \right)^{1/2}.
\end{aligned}
$$

Passing to the limit as $n \to +\infty$ we get for every $\delta > 0$ and $\alpha > 0$

$$
\lim_{n \to +\infty} \int_{D_\delta \setminus A_\alpha} |\nabla u_n - \nabla u|^2\, dx \leq \frac{2\varepsilon}{\delta} \left[\mathrm{meas}(D) C(\delta) \right]^{1/2},
$$

and, since $\varepsilon > 0$ is arbitrary, the proof is concluded. ∎

The compactness result above allows us to obtain an existence result for optimization problems on the class E_M.

Theorem 2.1.4 *Let* $f : D \times \mathbb{R} \times \mathbb{R}^N \to \mathbb{R}$ *be a bounded function which satisfies conditions* **A1**, **A2**. *Then the optimization problem*

$$(2.14) \qquad \min \left\{ \int_D f(x, u, \nabla u) \, dx \; : \; u \in E_M \right\}$$

admits a solution for every $M \geq 0$.

Proof Let (u_n) be a minimizing sequence for problem (2.14); by the argument used in the first part of Lemma 2.1.3, passing to subsequences we may assume that $u_n \to u$ strongly in $L^2(D)$ and weakly in $H^1(D_\delta)$ for every $\delta > 0$, for a suitable $u \in E_M$. Moreover, always by Lemma 2.1.3, for every $\alpha > 0$ there exists an open set $A_\alpha \subset D$ with meas$(A_\alpha) < \alpha$ and a subsequence (which we still denote by (u_n)) such that

$$\nabla u_n \to \nabla u \qquad \text{a.e. in } D \setminus A_\alpha.$$

We may now apply Fatou's lemma to $f(x, u_n, \nabla u_n)$ on $D \setminus A_\alpha$ and we obtain

$$(2.15) \qquad \begin{aligned} & \int_D f(x, u, \nabla u) \, dx \\ &= \int_{D \setminus A_\alpha} f(x, u, \nabla u) \, dx + \int_{A_\alpha} f(x, u, \nabla u) \, dx \\ &\leq \liminf_{n \to +\infty} \int_{D \setminus A_\alpha} f(x, u_n, \nabla u_n) \, dx + \int_{A_\alpha} f(x, u, \nabla u) \, dx \\ &\leq \liminf_{n \to +\infty} \int_D f(x, u_n, \nabla u_n) \, dx + C\alpha. \end{aligned}$$

Finally, by letting $\alpha \to 0$ we get that u is a solution of problem (2.14). ∎

Remark 2.1.5 A more careful inspection of the proof above shows that the result of theorem 2.1.4 still holds under the weaker growth assumption:

A3 there exist $p < 2$ and two functions $a(x,t)$, $b(x,t)$ from $D \times \mathbb{R}$ into $\mathbb{R}$, increasing in t, $a(\cdot, t) \in L^1_{loc}(D)$, $b(\cdot, t) \in L^{2/(2-p)}_{loc}(D)$ such that

$$0 \leq f(x, s, z) \leq a(x, |s|) + b(x, |s|)|z|^p \qquad \forall (x, s, z) \in D \times \mathbb{R} \times \mathbb{R}^N.$$

Other constraints than prescribing the maximal height M of the body are possible. For instance, in the case of convex bodies, we can prescribe a bound V on the volume of the body, so that we deal with the admissible class

$$C^V = \{u : D \to \mathbb{R} \; : \; u \text{ concave }, \; u \geq 0, \; \int_D u \, dx \leq V\}.$$

Alternatively, we can prescribe a bound S on the side surface of the body, so that the admissible class becomes

$$C(S) = \{u : D \to \mathbb{R} \; : \; u \text{ concave }, \; u \geq 0, \; \int_D \sqrt{1 + |\nabla u|^2} \, dx \leq S\}.$$

In both cases we have an existence result similar to the one of Theorem 2.1.1. Indeed, if u is concave its sup-norm can be estimated in terms of its integral, as it is easily seen by comparing the body itself with the cone of equal height:

$$V \geq \int_D u \, dx \geq \frac{(\sup u) \operatorname{meas}(D)}{N + 1}.$$

Then the volume class C^V is included in the height class C_M where $M = V(N+1)/\operatorname{meas}(D)$ and the corresponding compactness result follows from the one of Lemma 2.1.2.

The case of surface bound is similar: indeed, the sup-norm of a concave function can be estimated in terms of the surface of its graph, as it is easily seen by comparing again the body itself with the cone of equal height and by using Lemma 2.2.2:

$$S \geq \int_D \sqrt{1 + |\nabla u|^2} \, dx \geq \frac{(\sup u)\mathcal{H}^{N-1}(\partial D)}{N}.$$

Then the surface class $C(S)$ is included in the height class C_M where $M = SN/\mathcal{H}^{N-1}(\partial D)$ and the corresponding compactness result again follows from the one of Lemma 2.1.2.

2.2   Some necessary conditions of optimality

Coming back to the Newton problem of minimal resistance, it is interesting to note that all solutions (we shall see that there is not uniqueness of the solution) of (2.6) verify a necessary condition of optimality, given by the following result.

Theorem 2.2.1 *Let u be a solution of problem (2.6). Then for a.e. $x \in D$ we have that $|\nabla u|(x) \notin]0, 1[$.*

In the proof of Theorem 2.2.1 we shall use the following lemma.

Lemma 2.2.2 *Let A, B be two N-dimensional closed convex subsets of $\mathbb{R}^N$ with $A \subset B$. Then $\mathcal{H}^{N-1}(\partial A) \leq \mathcal{H}^{n-1}(\partial B)$ and equality holds if and only if $A = B$.*

Proof Let $P : \partial B \to \partial A$ be the projection on the closed convex set A, which maps every point of ∂B in the point of ∂A of least distance. It is well known (see for instance Brezis [30], Proposition V.3) that P is Lipschitz continuous with Lipschitz constant equal to 1. Therefore, by the general properties of Hausdorff measures (see for instance Rogers [173], Theorem 29), we obtain the inequality

$$\mathcal{H}^{N-1}(\partial A) = \mathcal{H}^{N-1}\big(P(\partial B)\big) \leq \mathcal{H}^{N-1}(\partial B)$$

which proves the desired inequality.
In order to conclude the proof, if by contradiction $\mathcal{H}^{N-1}(\partial A) = \mathcal{H}^{N-1}(\partial B)$ and $A \neq B$, we can find a hyperplane S tangent to A such that, denoting by S^+ the half space bounded by S and containing A, it is

$$B \setminus S^+ \neq \emptyset.$$

It is easy to see that $B \setminus S^+$ contains an open set, so that

$$
\begin{aligned}
(2.16) \qquad \mathcal{H}^{N-1}(\partial A) \ &\leq \mathcal{H}^{N-1}\big(\partial(B \cap S^+)\big) \\
&= \mathcal{H}^{N-1}(\partial B) + \mathcal{H}^{N-1}(\partial B \cap S) - \mathcal{H}^{N-1}(\partial B \setminus S^+) \\
&< \mathcal{H}^{N-1}(\partial B)
\end{aligned}
$$

which contradicts the assumption $H^{N-1}(\partial A) = H^{N-1}(\partial B)$ and achieves the proof. ∎

Proof [of Theorem 2.2.1] Let $u \in C_M$ be a solution of problem (2.6) and let v be defined as the infimum of M and of all tangent planes to the convex set $\{(x, y) \in D \times \mathbb{R} : 0 \leq y \leq u(x)\}$ having slope not belonging to $]0, 1[$. It is easy to see that $v \in C_M$, $v \geq u$ on D, $|\nabla v|(x) \notin]0, 1[$ for a.e. $x \in D$, and that on the set $\{v \neq u\}$ it is

$$|\nabla v| \in \{0, 1\} \qquad \text{and} \qquad |\nabla u| \in]0, 1[.$$

Consider now the function $\tilde{f} : \mathbb{R}^+ \to \mathbb{R}^+$ given by

(2.17)
$$\tilde{f}(t) = \begin{cases} 1 - t/2 & \text{if } 0 \le t \le 1 \\ 1/(1+t^2) & \text{if } t > 1 \end{cases}$$

and the functional

$$\tilde{F}(u) = \int_D \tilde{f}(|\nabla u|) \, dx.$$

The function $\tilde{f}$ is convex on $\mathbb{R}^+$ and we have

$$\tilde{f}(t) \le \frac{1}{1+t^2} \qquad \forall t \ge 0.$$

Therefore,

$$F(u) \ge \tilde{F}(u) = \int_{\{u=v\}} \tilde{f}(|\nabla u|) \, dx + \int_{\{u \ne v\}} \tilde{f}(|\nabla u|) \, dx.$$

Since $\nabla u = \nabla v$ a.e. on the set $\{u = v\}$, we get

(2.18)
$$\begin{aligned} F(u) \ & \ge \int_{\{u=v\}} \tilde{f}(|\nabla v|) \, dx + \int_{\{u \ne v\}} \tilde{f}(|\nabla u|) \, dx \\ & = \tilde{F}(v) + \int_{\{u \ne v\}} \left[\tilde{f}(|\nabla u|) - \tilde{f}(|\nabla v|) \right] \, dx. \end{aligned}$$

Since $|\nabla v| \notin]0, 1[$ on D we have

$$\tilde{f}(|\nabla v|) = f(|\nabla v|) \qquad \text{a.e. on } D;$$

moreover, since on $\{u \ne v\}$ it is $|\nabla v| \in \{0, 1\}$ and $|\nabla u| \in]0, 1[$, we have on $\{u \ne v\}$

$$\tilde{f}(|\nabla u|) = 1 - \frac{|\nabla u|}{2}, \qquad \tilde{f}(|\nabla v|) = 1 - \frac{|\nabla v|}{2}.$$

Therefore,

(2.19)
$$\begin{aligned} F(u) \ & \ge F(v) + \frac{1}{2} \int_{\{u \ne v\}} \left[|\nabla v| - |\nabla u| \right] \, dx \\ & = F(v) + \frac{1}{2} \int_D \left[|\nabla v| - |\nabla u| \right] \, dx. \end{aligned}$$

By the coarea formula we obtain

$$\int_\Omega \left[|\nabla v| - |\nabla u| \right] dx = \int_0^M \left[\mathcal{H}^{N-1}(\{v = t\}) - \mathcal{H}^{N-1}(\{u = t\}) \right] dt;$$

moreover, for every t the sets $\{u \geq t\}$ and $\{v \geq t\}$ are convex and

$$\{u \geq t\} \subset \{v \geq t\}.$$

Then, by Lemma 2.2.2 we get

$$F(v) \leq F(u)$$

and equality holds if and only if $u = v$. Therefore, $|\nabla u|$ must be outside the interval $]0, 1[$ and the proof is achieved. ∎

For a problem of the form (2.2) let u be a solution; we assume that in an open set ω the function u

i) is of class C^2;

ii) does not attain the maximal value M;

iii) is strictly concave in the sense that its Hessian matrix is positive definite.

Moreover, we assume that the integrand f is smooth. Then it is easy to see that for every smooth function ϕ with compact support in ω we have $u + \varepsilon\phi \in C_M$ for ε small enough. Thus we can perform the usual first variation calculation which leads to the Euler-Lagrange equation

$$- \operatorname{div} \left(f_z(x, u, \nabla u) \right) + f_s(x, u, \nabla u) = 0 \quad \text{in } \omega.$$

In the case of Newton functional this becomes

$$\operatorname{div} \left(\frac{\nabla u}{(1 + |\nabla u|^2)^2} \right) = 0 \quad \text{in } \omega.$$

We can also perform the second variation; this gives for every ϕ

$$\int_\omega \left[f_{zz}(x, u, \nabla u)\nabla\phi\nabla\phi + 2f_{sz}(x, u, \nabla u)\phi\nabla\phi + f_{ss}(x, u, \nabla u)\phi^2 \right] dx \geq 0.$$

In particular, for the Newton functional we obtain for every ϕ

$$(2.20) \qquad \int_\omega \frac{2}{(1+|\nabla u|^2)^3} \left(4(\nabla u \nabla \phi)^2 - (1+|\nabla u|^2)|\nabla \phi|^2\right) dx \geq 0.$$

Condition (2.20) gives, as a consequence, the following result.

Theorem 2.2.3 *Let D be a circle. Then an optimal solution of the Newton problem*

$$(2.21) \qquad \min\left\{ \int_D \frac{1}{1+|\nabla u|^2}\, dx \ : \ u \in C_M \right\}$$

cannot be radial.

Proof We follow the proof given in [33], assuming for simplicity $N = 2$. Let u be the optimal radial solution of the Newton problem computed in Section 1.3; we have seen that, outside a circle of radius r_0 where $u \equiv M$, the function u is smooth, strictly concave, and does not attain the maximal value M. Then, using in (2.20) a function ϕ of the form $\eta(r)\psi(\theta)$ with spt $\eta \subset]r_0, R[$ being R the radius of D, we obtain

$$\int_{r_0}^R r\, dr \int_0^{2\pi} \left[\frac{4|u'(r)\eta'(r)\psi(\theta)|^2}{(1+|u'(r)|^2)^3} - \frac{|\eta'(r)\psi(\theta)|^2 + |\eta(r)\psi'(\theta)|^2 r^{-2}}{(1+|u'(r)|^2)^2} \right] d\theta \geq 0.$$

Using $\psi(k\theta)$ instead of $\psi(\theta)$ the previous inequality becomes

$$\int_{r_0}^R r\, dr \int_0^{2\pi} \left[\frac{4|u'(r)\eta'(r)\psi(\theta)|^2}{(1+|u'(r)|^2)^3} - \frac{|\eta'(r)\psi(\theta)|^2 + k^2|\eta(r)\psi'(\theta)|^2 r^{-2}}{(1+|u'(r)|^2)^2} \right] d\theta \geq 0,$$

and the contradiction follows by taking $k \to +\infty$. ■

Remark 2.2.4 We may perform a similar computation for the integral

$$\int_D f(|\nabla u|)\, dx$$

and we find the second variation inequality

$$\int_\omega \frac{f'(|\nabla u|)}{|\nabla u|}|\nabla \phi|^2 + \left(\frac{f''(|\nabla u|)}{|\nabla u|^2} - \frac{f'(|\nabla u|)}{|\nabla u|^3} \right)(\nabla u \nabla \phi)^2\, dx \geq 0.$$

Assuming that the minimizer u is radial, the choice of ϕ as above leads to

$$\int_{r_1}^{r_2} \int_0^{2\pi} rf''(|u'|)|\eta'(r)\psi(\theta)|^2 + k^2\frac{f'(|u'|)}{r|u'|}|\eta(r)\psi'(\theta)|^2 \, dr \, d\theta \geq 0,$$

being $]r_1, r_2[$ an interval where u is smooth, strictly concave, and strictly less than M. Again, taking $k \to +\infty$ gives that the radial symmetry fails whenever $f'(|u'(r)|) < 0$ for some r, which implies the necessary condition of optimality

$$f'(|u'(r)|) \geq 0.$$

Remark 2.2.5 An immediate consequence of the nonradiality of the optimal Newton solutions is that problem (2.21) does not have a unique solution. In fact, rotating any nonradial solution u provides still a solution, as it is easy to verify, and therefore the number of solutions of problem (2.21) is infinite.

A more careful inspection of the proof of Theorem 2.2.3 allows us to obtain an additional necessary condition of optimality: all solutions of the Newton problem (2.21) must be "flat" in the sense specified by the following result (see [147]).

Theorem 2.2.6 *Let D be any convex domain and let u be a solution of the Newton problem (2.21). Assume that in an open set ω the function u is of class C^2 and does not touch the upper bound M. Then*

(2.22) $$\det D^2 \equiv 0 \quad \text{in } \omega.$$

Proof Let us fix a point $x_0 \in \omega$ and denote by a a unit vector orthogonal to $\nabla u(x_0)$. If (2.22) does not hold, then the second variation argument gives inequality (2.20) for every smooth function ϕ with support in ω. Take now

$$\phi(x) = \eta(x)\sin(ka \cdot x)$$

where the support of η is in a small neighbourhood of x_0 and k is large enough. We have

$$\nabla\phi(x) = \sin(ka \cdot x)\nabla\eta(x) + ka\cos(ka \cdot x)\eta(x)$$

so that, passing to the limit as $k \to +\infty$, we obtain

$$\int_\omega \frac{2\eta^2(x)}{(1+|\nabla u|^2)^3}\left(4(a \cdot \nabla u)^2 - (1+|\nabla u|^2)\right) dx \geq 0$$

for all η. As the support of η shrinks to x_0 this gives a contradiction, since $a \cdot \nabla u(x_0) = 0$. ∎

Remark 2.2.7 The result of Theorem 2.2.6 gives in another way that the solutions of the Newton problem in the case of D a disc cannot be radial. Moreover, the same argument can be repeated for functionals of the form $\int_D f(\nabla u)\, dx$. In this case we obtain that every minimizer u has to satisfy

$$f_{zz}\big(\nabla u(x_0)\big) \geq 0 \qquad \text{whenever } u \text{ is } C^2 \text{ around } x_0.$$

Finally, the flatness of solutions can be obtained also without assuming C^2 regularity, as it can be found in [147].

Remark 2.2.8 Another, more direct, proof of the nonradiality of the solutions of the Newton probem when D is a disc, has been found by P. Guasoni in [131]. In fact, if S is the segment joining the points $(-a, 0, M)$ and $(a, 0, M)$, the convex hull of $S \cup (D \times \{0\})$ can be seen as the hypograph of a function $u_{a,M} \in C_M$ which is graphically represented in the figure below.

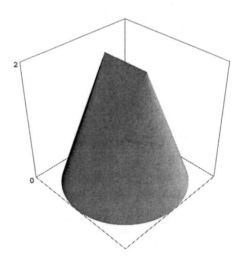

Figure 2.1: A "screwdriver" shape

If the number $a \in [0, R]$ is suitably chosen, the relative resistance of $u_{a,M}$ can be estimated and we obtain, after some calculations,

$$C_0(u_{a,M}) = \frac{1}{\pi R^2} \int_{B_{0,R}} \frac{1}{1 + |\nabla u_{a,M}|^2}\, dx \leq C(M/R)^{-2} + o\big((M/R)^{-2}\big)$$

as $M/R \to +\infty$. The constant C can be computed and we find $C < 27/32$ which shows (at least for large values of M/R) that the radial function of Section 1.3 cannot be a minimizer.

2.3 Optimization for boundary integrals

In this section we consider shape optimization problems of the form

$$(2.23) \qquad \min\left\{ \int_{\partial A} f(x, \nu(x)) \, d\mathcal{H}^{N-1} \ : \ A \in \mathcal{A} \right\}$$

where f is a nonnegative continuous function, ν is the normal unit vector exterior to A, and the class $\mathcal{A}$ of admissible domains is made of convex subsets of $\mathbb{R}^N$. This formulation allows us to consider convex bodies A which are not of Cartesian type, that is we do not need the admissible domains A are the subgraphs of concave functions u defined on a given convex set D.

The Newtonian resistance functional itself can be written in the form (2.23); in fact, for a Cartesian domain A given by the subgraph of a function u we have

$$\nu = \left(\frac{-\nabla u}{\sqrt{1 + |\nabla u|^2}}, \frac{1}{\sqrt{1 + |\nabla u|^2}} \right),$$

so that

$$\frac{1}{1 + |\nabla u|^2} = (\nu_N)^2.$$

Therefore, since changing the integration on D into an integral on ∂A provides an additional factor $(1 + |\nabla u|^2)^{-1/2} = \nu_N$, the Newtonian resistance functional takes the form

$$F(A) = \int_D \frac{1}{1 + |\nabla u|^2} \, dx = \int_{\text{graph } u} \nu_N^3(x) \, d\mathcal{H}^{N-1} = \int_{\partial A} \left(\nu_N^+(x) \right)^3 d\mathcal{H}^{N-1},$$

where the positive part in $\nu_N^+(x)$ is due to the fact that we do not want to take into account the lower horizontal part $\partial A \setminus \text{graph } u = D \times \{0\}$, on which $\nu_N < 0$. More generally, if a is the direction of the motion of the fluid stream, the Newtonian resistance has the form (2.23) with

$$f(x, \nu) = \left((a \cdot \nu)^+ \right)^3.$$

The admissible class we consider is

(2.24) $C_{K,Q} = \{A \text{ convex subset of } \mathbb{R}^N \; : \; K \subset A \subset Q\}$

where K and Q are two given compact subsets of $\mathbb{R}^N$. In the case on Newton's problem with prescribed height of Sections 1.3 and 2.1 we have

$$Q = \overline{D} \times [0, M], \qquad K = \overline{D} \times \{0\}.$$

The existence result we are going to prove is the following.

Theorem 2.3.1 *Let $f : \mathbb{R}^N \times S^{N-1} \to [0, +\infty]$ be a lower semicontinuous function and let K and Q be two given compact subsets of $\mathbb{R}^N$ such that the class $C_{K,Q}$ is nonempty. Then the minimum problem*

(2.25) $$\min\left\{ \int_{\partial A} f(x, \nu(x)) \, d\mathcal{H}^{N-1} \; : \; A \in C_{K,Q}\right\}$$

admits at least one solution.

We shall use several notions about measures, collected in the following definition.

Definition 2.3.2 *For every Borel measure μ on $\mathbb{R}^N$ with values in $\mathbb{R}^N$ we define the variation of μ as the nonnegative measure $|\mu|$ defined for every Borel subset B of $\mathbb{R}^N$ by*

$$|\mu|(B) = \sup\left\{ \sum_n |\mu(B_n)| \; : \; \cup_n B_n = B\right\}.$$

We denote by $\mathcal{M}$ the class of all measures μ such that $|\mu|(\mathbb{R}^N) < +\infty$, and for each $\mu \in \mathcal{M}$ we set
$$\|\mu\| = |\mu|(\mathbb{R}^N).$$

If $\mu \in \mathcal{M}$ the symbol ν_μ will denote the Radon-Nikodym derivative $d\mu/d|\mu|$, which is a μ-measurable function from $\mathbb{R}^N$ into S^{N-1}.
Finally we say that a sequence (μ_h) of measures in $\mathcal{M}$ converges in variation to μ if

$$\mu_h \to \mu \quad \text{weakly in } \mathcal{M} \qquad \text{and} \qquad \lim_{n \to +\infty} \|\mu_h\| = \|\mu\|.$$

The main tool we use in the proof of Theorem 2.3.1 is the following Reshetnyak result (see [170]) on functionals defined on measures.

Theorem 2.3.3 *Let $f : \mathbb{R}^N \times S^{N-1} \to \mathbb{R}$ be a bounded continuous function. Then the functional $F : \mathcal{M} \to \mathbb{R}$ defined by*

$$(2.26) \qquad F(\mu) = \int_{\mathbb{R}^N} f(x, \nu_\mu) \, d|\mu|$$

is continuous with respect to the convergence in variation.

Corollary 2.3.4 *If $f : \mathbb{R}^N \times S^{N-1} \to [0, +\infty]$ is lower semicontinuous, then the functional defined in (2.26) turns out to be lower semicontinuous with respect to the convergence in variation.*

Proof It is enough to approximate the function f by an increasing sequence (f_n) of bounded continuous functions, to apply to every functional

$$F_n(\mu) = \int_{\mathbb{R}^N} f_n(x, \nu_\mu) \, d|\mu|$$

the result of Theorem 2.3.3, and to pass to the supremum as $n \to +\infty$. ∎

The following Lemma will be also used.

Lemma 2.3.5 *Let A_n, A be bounded convex subsets of $\mathbb{R}^N$ with $A_n \to A$ in $L^1(\mathbb{R}^N)$. Then*

$$\lim_{n \to +\infty} \mathcal{H}^{N-1}(\partial A_n) = \mathcal{H}^{N-1}(\partial A).$$

Proof As A_n converges to A in $L^1(\mathbb{R}^N)$ it follows that

$$\forall \varepsilon > 0 \quad \exists n_\varepsilon : n > n_\varepsilon \quad \Rightarrow \quad A_n \subset A + B_{0,\varepsilon}.$$

Therefore, by Lemma 2.2.2, we obtain for $n > n_\varepsilon$,

$$\mathcal{H}^{N-1}(\partial A_n) \leq \mathcal{H}^{N-1}\big(\partial(A + B_{0,\varepsilon})\big)$$

so that

$$\limsup_{n \to +\infty} \mathcal{H}^{N-1}(\partial A_n) \leq \limsup_{\varepsilon \to 0^+} \mathcal{H}^{N-1}\big(\partial(A + B_{0,\varepsilon})\big) = \mathcal{H}^{N-1}(\partial A).$$

On the other hand, by the L^1 lower semicontinuity of the perimeter,

$$\liminf_{n\to+\infty} \mathcal{H}^{N-1}(\partial A_n) \geq \mathcal{H}^{N-1}(\partial A)$$

and the proof is complete.   ∎

Proof [of Theorem 2.3.1] It is convenient to restate the problem in terms of functionals depending on vector measures. To this aim, to every convex set $A \in C_{K,Q}$ we associate its characteristic function 1_A defined by

(2.27)
$$1_A(x) = \begin{cases} 1 & if x \in A \\ 0 & if x \notin A \end{cases}$$

and the distributional gradient $\nabla 1_A$ which is a vector measure of the class $\mathcal{M}$. It is well known that the measures $|\nabla 1_A|$ and $\mathcal{H}^{N-1}\llcorner\partial A$ coincide, so that the cost functional can be written in the form

$$\int_{\partial A} f\big(x,\nu(x)\big)\,d\mathcal{H}^{N-1} = \int_Q f\big(x,\nu_{\mu_A}(x)\big)\,d|\mu_A|$$

where we denoted by μ_A the measure $\nabla 1_A$. By the Reshetnyak Theorem 2.3.3 and the related Corollary 2.3.4 the functional above is lower semicontinuous with respect to the convergence in variation of the measures μ_A, so in order to apply the direct methods of the calculus of variations it remains to show that the class

$$M_{K,Q} = \big\{\mu \in \mathcal{M} \ : \ \mu = \nabla 1_A, \ A \in C_{K,Q}\big\}$$

is compact for the same convergence. Let (A_n) be a sequence of convex domains of $C_{K,Q}$; by Lemma 2.2.2 we have

$$\|\nabla 1_{A_n}\| = \mathcal{H}^{N-1}(\partial A_n) \leq \mathcal{H}^{N-1}(\partial \tilde{Q})$$

where $\tilde{Q}$ denotes the convex envelope of Q. Hence the sequence (1_{A_n}) is bounded in BV, so that we may assume, up to extracting a subsequence, it converges weakly* in BV to some function of the form 1_A. In particular we have $A_n \to A$ strongly in L^1, which implies that A is a convex domain of $C_{K,Q}$, and by Lemma 2.3.5

$$\lim_{n\to+\infty} \|\nabla 1_{A_n}\| = \lim_{n\to+\infty} \mathcal{H}^{N-1}(\partial A_n) = \mathcal{H}^{N-1}(\partial A) = \|\nabla 1_A\|,$$

which gives the required convergence in variation and concludes the proof.∎

Remark 2.3.6 All the arguments above work in a similar way if instead of the class $C_{K,Q}$ we work with a volume constraint and so with the admissible class

$$C^{V,Q} = \{A \text{ convex subset of } \mathbb{R}^N \ : \ A \subset Q, \ \text{meas}(A) \geq V\}.$$

Anther possible choice for the admissible class (see Buttazzo and Guasoni [69]) is obtained if also the section of the unknown domain A, with respect to a given hyperplane π, is involved in the optimization. We then have the class

$$S_{K,Q,m} = \{A \text{ convex subset of } \mathbb{R}^N \ : \ K \subset A \subset Q, \ \mathcal{H}^{N-1}(A \cap \pi) \geq m\}$$

for which all the previous analysis can be repeated.

2.4 Problems governed by PDE of higher order

In this section we deal with optimization problems on classes of convex domains, of a type different from the ones considered in Section 2.3. In particular, the class of admissible domains will be similar to the one of Section 2.3, that is

(2.28)
$$C_m(K,Q) = \{A \text{ convex subset of } \mathbb{R}^N \ : \ K \subset A \subset Q, \ \text{meas}(A) = m\}$$

where K and Q are two given compact subsets of $\mathbb{R}^N$. The cost functional, however, is of a different type and may involve PDE of higher order as a state equation. Problems of this type have been studied for instance in [180].

Let us start by introducing some useful notions about convex sets and by studying their properties. A natural topology on the class of convex sets is given by the so called Hausdorff distance.

Definition 2.4.1 *The Hausdorff distance between two closed sets A, B of $\mathbb{R}^N$ is defined by*

$$d(A,B) = \sup_{x \in A} d(x,B) \vee \sup_{x \in B} d(x,A)$$

where $d(x,E) = \inf\{|x - y| \ : \ y \in E\}$.

Remark 2.4.2 It is well known that the class of all closed subsets of a given compact set is compact with respect to the Hausdorff distance. Moreover, the convergence $A_n \to A$ induced by the Hausdorff distance is equivalent to the so called uniform convergence, which occurs if for every $\varepsilon > 0$ there exists n_ε such that

$$A_n \subset A + B_{0,\varepsilon} \quad \text{and} \quad A \subset A_n + B_{0,\varepsilon} \qquad \forall n \geq n_\varepsilon$$

being $B_{0,\varepsilon}$ the ball in R^N centered at the origin and of radius ε.

We summarize here below some of the properties of convex sets.

Proposition 2.4.3 *The following facts hold for convex sets.*

i) If $A \subset B$ then $\mathcal{H}^{N-1}(\partial A) \leq \mathcal{H}^{N-1}(\partial B)$;

ii) If $A_n \to A$ uniformly, then $A_n \to A$ in L^1, hence $\mathrm{meas}(A_n) \to \mathrm{meas}(A)$ and $\mathcal{H}^{N-1}(\partial A_n) \to \mathcal{H}^{N-1}(\partial A)$;

iii) $\mathrm{meas}(A) < \rho \mathcal{H}^{N-1}(\partial A)$ where ρ is the radius of the largest ball included in A.

Proof Statement i) is proved in Lemma 2.2.2. To prove statement ii) it is enough to notice that, by the definition of uniform convergence we have for every $\varepsilon > 0$

$$A_n \setminus A \subset \left(A + B_{0,\varepsilon}\right) \setminus A \qquad \text{for all } n \text{ large enough}$$

so that $\mathrm{meas}(A_n \setminus A) \leq C\varepsilon$ for a suitable constant C. Analogously we have $\mathrm{meas}(A \setminus A_n) \leq C\varepsilon$ which gives the L^1 convergence of A_n to A and the rest of the statement follows from Lemma 2.3.5. Finally, for the proof of statement iii) we refer to [166]. ∎

Proposition 2.4.4 *The class $C_m(K, Q)$ defined in (2.28) is compact for the uniform convergence.*

Proof Let (A_n) be a sequence in $C_m(K, Q)$; since all A_n are contained in the convex envelope $co(Q)$ of Q, by Proposition 2.4.3 i) we obtain

$$\mathcal{H}^{N-1}(\partial A_n) \leq \mathcal{H}^{N-1}(co(Q))$$

so that by Proposition 2.4.3 iii) we have that the largest ball included in A_n has a radius

$$\rho_n > m/\mathcal{H}^{N-1}(co(Q)) .$$

Therefore, possibly passing to a subsequence, that we still denote by (A_n), we may assume that there exists a ball $B_{x_0,\rho}$ with $\rho > 0$, which is contained in every A_n. Then the boundary ∂A_n can be described in the polar form

$$x - x_0 = r_n(\theta) \qquad x \in \partial A_n, \ \theta \in S^{N-1}.$$

Since $B_{x_0,\rho} \subset A_n \subset Q$ it is easy to see that the functions r_n have to be equi-Lipschitz continuous, so that by Ascoli-Arzelà theorem we may assume they converge uniformly to some function $r(\theta)$. This function describes the boundary of the limit set A by the polar form

$$x = r(\theta) \qquad x \in \partial A, \ \theta \in S^{N-1}.$$

Thus we have $A_n \to A$ uniformly; moreover it is easy to see that $A \in C_m(K, Q)$, which achieves the proof. ∎

Theorem 2.4.5 *Let* $J : C_m(K, Q) \to [0, +\infty]$ *be a cost functional which is lower semicontinuous with respect to the uniform convergence; then the optimization problem*

$$\min \big\{ J(A) \ : \ A \in C_m(K, Q) \big\}$$

admits at least a solution.

Proof The proof is a straightforward consequence of Proposition 2.4.4 and of the direct methods of the calculus of variations. ∎

As an application of the previous result we present here two examples taken from [180] where the related optimization problems involve PDE of higher order.

In the first example we consider an elliptic operator L of order 2ℓ, of the form

(2.29) $$Lu = \sum_{|\alpha|,|\beta|=\ell} (-1)^\ell D^\alpha \big(a_{\alpha,\beta}(x) D^\beta u \big)$$

where the coefficients $a_{\alpha,\beta}$ are bounded and satisfy the ellipticity condition

$$c_0 \sum_{|\alpha|=\ell} \xi^{2\alpha} \leq \sum_{|\alpha|,|\beta|=\ell} a_{\alpha,\beta}(x) \xi^\alpha \xi^\beta$$

for every $\xi \in \mathbb{R}^N$, where c_0 is a positive constant. For every $A \in C_m(K, Q)$ we denote by $\lambda_j(A)$ the j-th eigenvalue of L, counted with its multiplicity, on the Sobolev space $H_0^\ell(A)$, and by $e_{j,A}$ the corresponding eigenfunction which satisfies the equation

$$(2.30) \qquad \begin{cases} Lu = \lambda_j(A)u & \text{in } A \\ \quad u \in H_0^\ell(A). \end{cases}$$

It is well known that $\lambda_j(A)$ admits the following variational characterization:

$$\lambda_j(A) = \max_{H \in \Sigma_j} \min \left\{ \langle Lu, u \rangle \; : \; u \in H, \; \int_A u^2 \, dx = 1 \right\}$$

where Σ_j is the class of all linear subspaces of $H_0^\ell(A)$ of dimension j. Therefore it is easy to prove that all $\lambda_j(A)$ are monotone decreasing as functions of the domain A, with respect to the set inclusion. Moreover, in terms of eigenfunctions we also have

$$(2.31) \qquad \lambda_j(A) = \min \Big\{ \; \langle Lu, u \rangle \; : \; u \in H_0^\ell(A), \; \int_A u^2 \, dx = 1, \\ \int_A u e_{i,A} \, dx = 0 \text{ for } i < j \Big\}.$$

By Proposition 2.4.3 iii) we have

$$\lambda_j\big(co(Q)\big) \leq \lambda_j(A) \leq \lambda_j\big(B_{x,\rho}\big)$$

where $B_{x,\rho}$ denotes the largest ball included in A. Since ρ is bounded from below, the previous inequality shows that for every integer j the quantity $\lambda_j(A)$ is bounded when A varies in $C_m(K, Q)$.

Proposition 2.4.6 *For every integer j the mapping $\lambda_j : C_m(K, Q) \to \mathbb{R}$ is continuous for the uniform convergence.*

Proof Fix an integer j and take a sequence (A_n) in $C_m(K, Q)$ converging to A uniformly. Up to extracting a subsequence, thanks to Proposition 2.4.3, we may assume that all A_n and A contain a ball of radius ρ centered in a point that, without loss of generality, we may assume to be the origin. Moreover

$A_n \to A$ in L^1. Then by Remark 2.4.2 and by the monotonicity of λ_j, for every $\varepsilon > 0$ we have for n large enough

(2.32)
$$\begin{cases} \lambda_j(A) \geq \lambda_j(A_n + B_{0,\varepsilon}) \geq \lambda_j\big((1 + c\varepsilon)A_n\big) \\ \lambda_j(A_n) \geq \lambda_k\big(A + B_{0,\varepsilon}\big) \geq \lambda_j\big((1 + c\varepsilon)A\big) \end{cases}$$

where the constant $c > 0$ can be taken independent of n and ε. It is now easy, by repeating the arguments already seen in Section 1.4, and by using (2.31), to show that $\lambda_j(A_n) \to \lambda_j(A)$ and that the corresponding eigenfunctions $e_{j,A_n} \to e_{j,A}$ strongly in $H_0^\ell(\mathbb{R}^N)$. ∎

Let us consider now a cost functional of the form

$$F(A) = \Phi\big(\Lambda(A)\big)$$

where $\Lambda(A)$ denotes the whole spectrum of the operator L over $H_0^\ell(A)$. We assume that the function Φ is lower semicontinuous, in the sense that

$$\Phi(\Lambda) \leq \liminf_{n \to +\infty} \Phi(\Lambda_n) \qquad \text{whenever } \Lambda_n \to \Lambda$$

where the convergence $\Lambda_n \to \Lambda$ is defined by

$$\Lambda_n \to \Lambda \quad \Longleftrightarrow \quad \lambda_{j,n} \to \lambda_j \quad \forall j = 1, \dots.$$

In particular, if Φ depends only on a finite number M of variables, then the lower semicontinuity above reduces to the usual lower semicontinuity in $\mathbb{R}^M$.

Theorem 2.4.7 *Let Φ be lower semicontinuous in the sense above. Then the optimization problem*

$$\min\big\{\Phi\big(\Lambda(A)\big) \ : \ A \in C_m(K, Q)\big\}$$

admits at least a solution.

Proof It is enough to apply the direct methods of the calculus of variations, taking into account the results of Proposition 2.4.6 and of Proposition 2.4.4. ∎

In the second example we consider again an operator of the form 2.29 and a cost functionals

$$F(A) = \int_{\mathbb{R}^N} j(x, u_A, \nabla u_A, \ldots, D^\ell u_A) \, dx$$

where we denoted by u_A the solution of

(2.33)
$$\begin{cases} Lu = f & \text{in } A \\ u \in H_0^\ell(A) \end{cases}$$

being f a given function in $L^2(\mathbb{R}^N)$, or more generally in $H^{-\ell}(\mathbb{R}^N)$.

Theorem 2.4.8 *Assume that j is a nonnegative Borel function such that $j(x, \cdot, \ldots, \cdot)$ is lower semicontinuous. Then the optimization problem*

$$\min\left\{F(A) \; : \; A \in C_m(K, Q)\right\}$$

admits at least a solution.

Proof It is enough to repeat the arguments used in the proof of Theorem 2.4.7, noticing that, as before, we have $u_{A_n} \to u_A$ strongly in $H_0^\ell(\mathbb{R}^N)$ whenever $A_n \to A$ uniformly. ∎

Chapter 3

Optimal control problems: a general scheme

Optimal control problems are minimum problems which describe the behaviour of systems that can be modified by the action of an operator. Many problems in applied sciences can be modelized by means of optimal control problems. Two kinds of variables (or sets of variables) are then involved: one of them describes the state of the system and cannot be modified directly by the operator, it is called the *state variable*; the second one, on the contrary, is under the direct control of the operator that may choose its strategy among a given set of admissible ones, it is called the *control variable*.

The operator is allowed to modify the state of the system indirectly, acting directly on control variables; only these may act on the system, through a link control-state, usually called *state equation* . Finally, the operator, acting directly on controls and indirectly on states through the state equation, must achieve a goal usually written as a minimization of a functional, which depends on the control that has been chosen as well as on the corresponding state, the so-called *cost functional*.

Driving a car is a typical example of an optimal control problem: the driver may only act directly on controls which are in this case the accelerator, the brakes, and the steering-wheel; the state of the car is on the contrary described by its position and velocity which, of course, depend on the controls chosen by the driver, but are not directly controlled by him. The state equations are the usual equations of mechanics which, to a given choice of acceleration and steering angle, associate the position and velocity of the car, also taking into account the specifications of the engine (technological constraints, nonlinear behaviours, ...). Finally, the driver wants to achieve a goal, for instance to minimize the total fuel consumption to run along a given path. Then we have an optimal control problem, where the driver has to choose the best driving strategy to minimize the cost functional, which is in this case the total fuel consumption.

According to what said above the ingredients of an optimal control problem are:

i) a space of states Y;

ii) a set of controls U;

iii) a the set $\mathcal{A}$ of *admissible pairs*, that is a subset of pairs $(u, y) \in U \times Y$ such that y is linked to u through the state equation;

iv) a cost functional $J : U \times Y \to \overline{\mathbb{R}}$.

The optimal control problem then takes the form of a minimization problem written as:

$$\min \big\{ J(u, y) \ : \ (u, y) \in \mathcal{A} \big\}.$$

We are specially interested in the study of shape optimization problems, where the control variable runs over classes of domains. For this reason we have to consider a framework general enough to include cases when the control variable does not belong to a space with a linear topological structure. On the contrary, taking the state variable as an element of a space of functions (a Sobolev space, a space of functions with bounded variation, ...) seems to fit most of the known situations from the applications. This is why in the list i) – iv) above we stressed the difference between the *space* Y and the *set* U.

The choice of a topology on Y and U is a very important matter when dealing with the question of existence of solutions to an optimal control problem. This is related to the use of direct methods of the Calculus of Variations, which require, for the problem under consideration, suitable lower semicontinuity and compactness assumptions.

In several cases of shape optimization problems it is known that an optimal solution does not exist; therefore minimizing sequences of domains cannot converge to an admissible domain, in any sense which preserves the lower semicontinuity of the cost functional. In order to study the asymptotic behaviour of minimizing sequences we shall endow U with an *ad hoc* topology, mainly depending on the state equation considered, and limits of minimizing sequences will be seen as optimal *relaxed* solutions which then turn out to belong to a larger space. In this chapter we give a rather general way of constructing this larger space of *relaxed controls*. Due to the great generality of our framework, the relaxed controls will be characterized simply as the elements of a Cauchy completion of a metric space; of course, when dealing with a more specific optimization problem, a more precise characterization will be needed: in the rest of these notes we shall see some relevant examples where this can be done.

3.1 A topological framework for general optimization problems

In this section we consider an abstract optimal control problem of the form

$$(3.1) \qquad \min \big\{ J(u,y) \; : \; (u,y) \in \mathcal{A} \big\}$$

where Y is the space of states, U is the set of controls, $J : U \times Y \to \overline{\mathbb{R}}$ is the cost functional, and $\mathcal{A} \subset U \times Y$ is the set of admissible pairs, determined in the applications by a state equation. We assume that Y is a separable metric space, while the controls vary in a set U with no topological structure a priori given. As already remarked in the introduction of Chapter 3 this happens in some quite important situations like shape optimization problems where the set of controls is given by suitable classes of admissible domains. To handle this situation it is convenient to write the set $\mathcal{A}$ of admissible pairs in the form

$$(3.2) \qquad \mathcal{A} = \big\{ (u,y) \in U \times Y \; : \; y \in \operatorname{argmin} G(u,\cdot) \big\},$$

where $G : U \times Y \to \overline{\mathbb{R}}$ is a given functional and where $\operatorname{argmin} G(u,\cdot)$ denotes the set of all minimum points of $G(u,\cdot)$. In the case $G(u,\cdot)$ is an integral functional of the calculus of variations whose integrand depends on the control u, its Euler-Lagrange equation provides the differential state equation. We shall call G the *state functional*. It is worth noticing that the set $\mathcal{A}$ can be always written in the form (3.2) by choosing

$$(3.3) \qquad G(u,y) = \chi_{\mathcal{A}}(u,y) = \begin{cases} 0 & \text{if } (u,y) \in \mathcal{A} \\ +\infty & \text{otherwise.} \end{cases}$$

Therefore, the optimal control problem (3.1) can be written in the form

(3.4) $$\min \left\{ J(u, y) \ : \ y \in \operatorname{argmin} G(u, \cdot) \right\}$$

For instance, a state equation like

(3.5) $$\begin{cases} -\Delta y = f & \text{in } A \\ y \in H_0^1(A) \end{cases}$$

is provided by the state functional

$$G(A, y) = \int_{\mathbb{R}^N} |\nabla y|^2 \, dx - \langle f, y \rangle + \chi_{H_0^1(A)}(y) \,,$$

where the states vary in the Sobolev space $H^1(\mathbb{R}^N)$ and the controls A vary in a class of domains.

Let us notice that in the applications the space Y of states is usually a separable reflexive Banach space of functions endowed with its weak topology (or the dual of a separable Banach space, endowed with its weak* topology), which is not, unless it is finite dimensional, metrizable. However, thanks to some growth assumptions on the cost functional J, we may often reduce to work on a bounded subset of Y which, as it is well known, is metrizable.

We shall endow U with a topology which is constructed by means of the functional G: the natural topology on U that takes into account the convergence of minimizers of G is the one related to the Γ-convergence of the mappings $G(u, \cdot)$ and will then be denoted by γ-convergence. Clearly, as soon as the convergence of controls implies the convergence of the associated states, it would be enough to have the compactness of minimizing sequences in U and the lower semicontinuity of the cost functional J in $U \times Y$ to obtain, always thanks to direct methods of the calculus of variation, the existence of an optimal pair (u, y). The lower semicontinuity of the cost functional J is not a very restrictive assumption: indeed in several cases J depends only on the state y in a continuous, or even more regular, way. On the contrary, the compactness of the set U, once endowed with the γ-convergence, is a rather severe requirement that in many cases does not occur: γ-limits of minimizing sequences may not belong to U. We will then construct a larger space of *relaxed controls* which is γ-compact so that the existence of an optimal relaxed solution will follow straightforward.

3.2 A quick survey on Γ-convergence theory

We recall here briefly the definition and the main properties of Γ-convergence. We do not want here to enter into the details of that theory, but only to use it in order to characterize the relaxed optimal control problem; we refer for all details to the book by Dal Maso [89]. In what follows Y denotes a separable metric space, endowed with a distance d.

Definition 3.2.1 *Given a sequence (G_n) of functionals from Y into $\overline{\mathbb{R}}$ we say that (G_n) Γ-converges to a functional G if for every $y \in Y$ we have:*

i) $\forall y_n \to y \quad G(y) \leq \liminf_{n \to +\infty} G_n(y_n);$

ii) $\exists y_n \to y \quad G(y) \geq \limsup_{n \to +\infty} G_n(y_n).$

We list here below the main properties of Γ-convergence.

• *Lower semicontinuity.* Every Γ-limit is lower semicontinuous on Y.

• *Convergence of minima.* If (G_n) Γ-converges to G and is equi-coercive on Y, that is for every $t \in \mathbb{R}$ there exists a compact set $K_t \subset Y$ such that

$$\{G_n \leq t\} \subset K_t \qquad \forall n \in \mathbb{N},$$

then G is coercive too and so it attains its minimum on Y. We have

$$\min G = \lim_{n \to +\infty} \left[\inf G_n\right].$$

• *Convergence of minimizers.* Let (G_n) be an equi-coercive sequence of functionals on Y which Γ-converges to a functional G. If $y_n \in \operatorname{argmin} G_n$ is a sequence with $y_n \to y$ in Y we have $y \in \operatorname{argmin} G$. Moreover, if G is not identically $+\infty$ and if $y_n \in \operatorname{argmin} G_n$, then there exists a subsequence of (y_n) which converges to an element of $\operatorname{argmin} G$. In particular, if G has a unique minimum point y on Y, then every sequence $y_n \in \operatorname{argmin} G_n$ converges to y in Y.

It is interesting to notice (see Proposition 7.7 in [89]) that a sequence (G_n) of functionals is equi-coercive in Y if and only if there exists a lower semicontinuous coercive function $\Psi : Y \to \overline{\mathbb{R}}$ such that $G_n \geq \Psi$ for all $n \in \mathbb{N}$.

• *Compactness.* From every sequence (G_n) of functionals on Y it is possible to extract a subsequence Γ-converging to a functional G on Y.

• *Metrizability.* The Γ-convergence, considered on the family $S(Y)$ of all lower semicontinuous functions on Y, does not come from a topology, unless the space Y is locally compact, which never occurs in the infinite dimensional case. However, if instead of considering the whole family $S(Y)$, we take the smaller classes

$$S_\Psi(Y) = \left\{ G : Y \to \overline{\mathbb{R}} \ : \ G \text{ l.s.c.}, \ G \geq \Psi \right\}$$

where $\psi : Y \to \overline{\mathbb{R}}$ is lower semicontinuous and coercive (and nonnegative, for simplicity), then the Γ-convergence on $S_\Psi(Y)$ is metrizable. More precisely, it turns out to be equivalent to the convergence associated to the distance

$$d_\Gamma(F, G) = \sum_{i,j=1}^{\infty} 2^{-i-j} \big| \arctan \big(F_j(y_i) \big) - \arctan \big(G_j(y_i) \big) \big|$$

where (y_i) is a dense sequence in Y and H_j denotes the Moreau-Yosida transforms of a functional H, defined by:

$$H_j(y) = \inf \big\{ H(x) + jd(x, y) \ : \ x \in Y \big\}.$$

According to the compactness property seen above, the family $S_\Psi(Y)$ endowed with the distance d_Γ turns out to be a compact metric space.

3.3   The topology of γ-convergence for control variables

We are now in a position to introduce a "natural" topology on the set U of control variables appearing in the general framework considered in Section 3.1.

Definition 3.3.1 *We say that $u_n \to u$ in U if the associated state functionals $G(u_n, \cdot)$ Γ-converge to $G(u, \cdot)$ in Y. This convergence on U will be called γ-convergence.*

We shall always assume in the following that the state functional G satisfies the following properties:

• for every $u \in U$ the function $G(u, \cdot)$ is lower semicontinuous in the space Y;

- G is equi-coercive in the sense that there exists a coercive lower semi-continuous functional $\Psi : Y \to \overline{\mathbb{R}}$ such that

$$G(u, y) \geq \Psi(y) \qquad \forall u \in U, \ \forall y \in Y.$$

- the mapping $\Gamma_G : U \to S_\Psi(Y)$ defined by $\Gamma_G(u) = G(u, \cdot)$ is one-to-one. Otherwise, we may always reduce the space U to a smaller space which verifies this property.

Remark 3.3.2 By the assumptions above we have, in particular, that for every $u \in U$ the set $\operatorname{argmin} G(u, \cdot)$ is nonempty. Moreover, according to the metrizability property of the Γ-convergence seen in Section 3.2, the γ-convergence on U is metrizable, and the mapping Γ_G is an isometry. However, even if $S_\Psi(Y)$ with the Γ-convergence is a compact metric space, in general U with the γ-convergence may be not compact. Indeed, a sequence $G(u_n, \cdot)$ of functionals may Γ-converge to a functional F, but this limit functional does not need to be of the form $G(u, \cdot)$ for some $u \in U$. This is why in many situations the existence of optimizers may fail and it is necessary to enlarge by relaxation the class of admissible controls U.

3.4 A general definition of relaxed controls

In this section we give the definition of relaxed controls in a rather general framework; the definition is given in the abstract scheme introduced in Section 3.1.

Definition 3.4.1 *The class $\hat{U}$ is defined as the completion of the metric space U endowed with the γ-convergence. The elements of $\hat{U}$ will be called relaxed controls and we still continue to denote by γ the convergence on $\hat{U}$.*

In order to define the relaxed optimal control problem associated to (3.1), (3.2) we have to introduce the relaxed cost functional $\hat{J}$ as well as the relaxed state functional $\hat{G}$. For every $\hat{u} \in \hat{U}$ we set

$$\hat{G}(\hat{u}, \cdot) = \Gamma \lim_{u \to \hat{u}} G(u, \cdot).$$

In other words, we define the mapping $\hat{\Gamma}_G : \hat{U} \to S_\psi(Y)$ as the unique isometry which extends Γ_G; more precisely,

$$\hat{\Gamma}_G(\hat{u}) = \Gamma \lim_{n \to +\infty} \Gamma_G(u_n),$$

where (u_n) is any sequence γ-converging to $\hat{u}$. Therefore we have $\hat{G} : \hat{U} \times Y \to \mathbb{R}$ defined by

$$\hat{G}(\hat{u}, \cdot) = \hat{\Gamma}_G(\hat{u}) \qquad \forall \hat{u} \in \hat{U}$$

and we have

$$\hat{u}_n \to \hat{u} \text{ in } \hat{U} \quad \Longleftrightarrow \quad \Gamma \lim_{n \to +\infty} \hat{G}(\hat{u}_n, \cdot) = \hat{G}(\hat{u}, \cdot).$$

Proposition 3.4.2 *The metric space $\hat{U}$ is compact with respect to the γ-convergence.*

Proof Since $\hat{\Gamma}_G$ is an isometry and $\hat{U}$ is complete, $\hat{\Gamma}_G(\hat{U})$ is a complete subspace of the compact space $S_\psi(Y)$, so that $\hat{\Gamma}_G(\hat{U})$ is compact. Hence, using again the fact that $\hat{\Gamma}_G$ is an isometry, we get that $\hat{U}$ is compact too. ∎

The definition of the relaxed state functional allows us to define the relaxed state equation, linking a relaxed control $\hat{u} \in \hat{U}$ to a state $y \in Y$, which reads now

$$y \in \operatorname{argmin} \hat{G}(\hat{u}, \cdot).$$

The relaxed cost functional $\hat{J}$ is defined in a similar way. Take a pair $(\hat{u}, y)$ which verifies the state equation, i.e. such that $y \in \operatorname{argmin} \hat{G}(\hat{u}, \cdot)$; then we set

$$\hat{J}(\hat{u}, y) = \inf \left\{ \liminf_{n \to +\infty} J(u_n, y_n) \ : \ u_n \to \hat{u} \text{ in } \hat{U}, \ y_n \to y \text{ in } Y, \right.$$
$$\left. y_n \in \operatorname{argmin} G(u_n, \cdot) \right\}.$$

Therefore the relaxed optimal control problem can be written in the form

$$(3.6) \qquad \min \left\{ \hat{J}(\hat{u}, y) \ : \ \hat{u} \in \hat{U}, \ y \in Y, \ y \in \operatorname{argmin} \hat{G}(\hat{u}, \cdot) \right\}.$$

In several situations the cost functional J depends only on the state y and is continuous on Y; in this case it is easy to see that $\hat{J} = J$ so that the relaxed optimal control problem has the simpler form

$$(3.7) \qquad \min \left\{ J(y) \ : \ \hat{u} \in \hat{U}, \ y \in Y, \ y \in \operatorname{argmin} \hat{G}(\hat{u}, \cdot) \right\}.$$

By the definition of relaxed control problem and by Proposition 3.4.2 we obtain immediately the following existence result.

Theorem 3.4.3 *Under the assumptions above the relaxed problem (3.6) admits at least a solution $(\hat{u}, y) \in \hat{U} \times Y$. Moreover, the infimum of the original problem given by (3.1) and (3.2) coincides with the minimum of the relaxed problem (3.6). Finally, if (u_n, y_n) is a minimizing sequence for the original problem, then there exists a subsequence converging in $\hat{U} \times Y$ to a solution $(\hat{u}, y)$ of the relaxed problem.*

Remark 3.4.4 On the one hand the result above gives the existence of an optimal pair $(\hat{u}, y)$ for a problem "*close*" to the original one; on the other hand the solution $\hat{u}$ belongs to a larger space and is only characterized as an element of an abstract topological completion, hence as an equivalence class of Cauchy sequences of the original control set U with respect to a quite involved distance function. In order to obtain further properties about the asymptotic behaviour of minimizing sequences is then necessary, in concrete cases, to give a more explicit characterization of the space of relaxed controls $\hat{U}$.

3.5 Optimal control problems governed by ODE

In this section we consider optimal control problems where the control variable varies in a space of functions. For simplicity we consider the case of problems where the state and the control variables are functions of one real variable; therefore the state equation will be an ordinary differential equation.

Example 3.5.1 A car has to go from a point A to a point B (for simplicity assume along a straight line) in a given time T. Setting

(3.8)
$$\begin{aligned}
y(t) &\quad : \text{the position of the car at the time } t \\
u(t) &\quad : \text{the acceleration we can give to the car}
\end{aligned}$$

we have the state equation

$$y'' = u .$$

In this problem the position $y(t)$ plays the role of state variable and the acceleration $u(t)$ is the control variable; in this case we can control the acceleration but not the speed and the position: they are given indirectly by the state equation $y'' = u$. We can assume further constraints on the control, like $|u| \leq 1$ ($u = -1$ representing the maximum action of brakes, $u = 1$ the maximum acceleration).

If we take as a cost functional the total fuel consumption, we have to consider that this consumption may depend on several variables, like for instance:

- u (how much you push the accelerator)

- y (if we are going up on a hill or going down)

- y' (higher is the speed higher is the consumption)

- t (on different hours of the day the consumption may be different).

Then the optimal control problem is given by the minimization of the functional

$$J(u, y) = \int_0^T f(t, y, y', u)\, dt$$

where the function f takes into accounts the variables above, with conditions

$$|u| \leq 1, \ y'' = u, \ y(0) = A, \ y(T) = B, \ y'(0) = 0.$$

Remark that an optimal solution is given by a pair (u, y).

   In the following we want to derive some simple conditions for the existence of a solution.

Lemma 3.5.2 *Assume that for $n \in \mathbb{N} \cup \{\infty\}$ the functions $g_n : [0, T] \times \mathbb{R}^N \to \mathbb{R}^N$ are measurable in t and equi-Lipschitz continuous in s, i.e.*

$$\exists L > 0 : |g_n(t, s_2) - g_n(t, s_1)| \leq L|s_2 - s_1|$$

for every $s_1, s_2 \in \mathbb{R}^N$, $t \in [0, T]$, $n \in \mathbb{N} \cup \{\infty\}$. Assume further that

$$|g_n(t, 0)| \leq M$$

and fix initial data $\xi_n \in \mathbb{R}^N$. If for all $n \in \mathbb{N} \cup \{\infty\}$ we denote by y_n the unique solutions of the differential equations

(3.9) $$\begin{cases} y'_n = g_n(t, y_n) & \text{in } [0, T] \\ y_n(0) = \xi_n, \end{cases}$$

then the conditions $\xi_n \to \xi_\infty$ and

$$g_n(\cdot, s) \to g_\infty(\cdot, s) \qquad \text{weakly in } L^1 \qquad \forall s \in \mathbb{R}^N$$

imply that $y_n \to y_\infty$ uniformly as $n \to +\infty$.

Proof It is convenient to write the differential equations in the integral form

(3.10)
$$y_n(t) = \xi_n + \int_0^t g_n(\tau, y_n(\tau)) \, d\tau$$
$$y_\infty(t) = \xi_\infty + \int_0^t g_\infty(\tau, y_\infty(\tau)) \, d\tau.$$

Take now piecewise constant functions y_ε such that $\|y_\varepsilon - y_\infty\|_{L^\infty} < \varepsilon$. Then we have

$$|y_n(t) - y_\infty(t)| \leq |\xi_n - \xi_\infty| + \left| \int_0^t g_n(\tau, y_n) \, d\tau - \int_0^t g_\infty(\tau, y_\infty) \, d\tau \right|$$

$$\leq |\xi_n - \xi_\infty| + \int_0^t |g_n(\tau, y_n) - g_n(\tau, y_\varepsilon)| \, d\tau$$

$$+ \left| \int_0^t g_n(\tau, y_\varepsilon) - g_\infty(\tau, y_\varepsilon) \, d\tau \right| + \int_0^t |g_\infty(\tau, y_\varepsilon) - g_\infty(\tau, y_\infty)| \, d\tau$$

$$\leq |\xi_n - \xi_\infty| + \int_0^t L|y_n - y_\varepsilon| \, d\tau + \left| \int_0^t g_n(\tau, y_\varepsilon) - g_\infty(\tau, y_\varepsilon) \, d\tau \right|$$

$$+ L\|y_\varepsilon - y_\infty\|$$

$$\leq |\xi_n - \xi_\infty| + L \int_0^t |y_n - y_\infty| \, d\tau + \left| \int_0^t g_n(\tau, y_\varepsilon) - g_\infty(\tau, y_\varepsilon) \, d\tau \right| + C\varepsilon.$$

Since y_ε is piecewise constant we have

$$\left| \int_0^t g_n(\tau, y_\varepsilon) - g_\infty(\tau, y_\varepsilon) \, d\tau \right| \to 0 \quad \text{uniformly as } n \to +\infty$$

so that

$$|y_n(t) - y_\infty(t)| \leq L \int_0^t |y_n(\tau) - y_\infty(\tau)| \, d\tau + \omega(n, \varepsilon)$$

where $\omega(n, \varepsilon) \to C\varepsilon$ as $n \to +\infty$. Applying now the Gronwall's lemma we obtain

$$|y_n(t) - y_\infty(t)| \leq \omega(n, \varepsilon) \exp \left(\int_0^1 L(\tau) \, d\tau \right).$$

Thus for a suitable constant C

$$\|y_n - y_\infty\| \leq C\omega(n, \varepsilon)$$

and, as ε was arbitrary, we get that $y_n \to y_\infty$ uniformly. ∎

Remark 3.5.3 The result of lemma above holds as well if the constant L depends on t in an integrable way.

By using the result of Lemma 3.5.2 we will prove an existence result for optimal control problems governed by equations of the form

$$y' = a(t, y) + b(t, y)u \, ,$$

with $a : [0, T] \times \mathbb{R}^N \to \mathbb{R}^N$ and $b : [0, T] \times \mathbb{R}^N \to \mathbb{R}^{Nm}$ measurable in t, Lipschitz continuous, and bounded at $y = 0$.
Let $f(t, s, z)$ be a Borel function such that

- $f \geq 0$;

- f is l.s.c in (s, z);

- f is convex in z.

Proposition 3.5.4 *Under the assumptions above, the functional*

$$F : L^p([0, T]; \mathbb{R}^m) \times W^{1,1}([0, T]; \mathbb{R}^N) \to [0, +\infty]$$

defined as

$$F(u, y) = \int_0^T f(t, y, u) \, dt + \chi_{\{y'=a(t,y)+b(t,y)u, \ y(0)=y_0\}}$$

is sequentially lower semicontinuous with respect to the $w - L^p \times w - W^{1,1}$ topology.

Proof Assume $u_n \to u$ weakly in L^p and $y_n \to y$ weakly in $W^{1,1}$. We can assume that for $n \in \mathbb{N}$

$$y'_n = a(t, y_n) + b(t, y_n)u_n \, , \qquad y_n(0) = y_0.$$

Defining

$$(3.11) \qquad \begin{aligned} g_n(t, s) &= a(t, s) + b(t, s)u_n(t) \\ g_\infty(t, s) &= a(t, s) + b(t, s)u(t) \, , \end{aligned}$$

the assumptions of Lemma 3.5.2 are fulfilled, hence we have

$$y' = a(t, y) + b(t, y)u \, , \qquad y(0) = y_0.$$

Therefore
$$F(u, y) = \int_0^T f(t, y, u) \, dt$$
and the lower semicontinuity follows from the general lower semicontinuity result for integral functionals. ∎

It remains to show the coercivity of the functional F. For this we need some assumptions:

- if $p \in]1, +\infty[$ we assume that there exist $\alpha > 0$ and $\gamma \in L^1$ such that
$$f(t, s, z) \geq \alpha |z|^p - \gamma(t) \, ;$$

- if $p = 1$ we assume that there exist a superlinear function ϕ and $\gamma \in L^1$ such that
$$f(t, s, z) \geq \phi(|z|) - \gamma(t) \, ;$$

- if $p = \infty$ we assume that there exist a positive number R and $\gamma \in L^1$ such that
$$f(t, s, z) \geq \chi_{\{|z| \leq R\}} - \gamma(t) \, .$$

Proposition 3.5.5 *Under the assumptions above the functional F is coercive with respect to the $w - L^p \times w - W^{1,1}$ topology.*

Proof Let $F(u_n, y_n) \leq c$. Hence for a subsequence we have $u_n \to u$ weakly in L^p (weakly* in L^∞ if $p = \infty$) and
$$y_n' = a(t, y_n) + b(t, y_n) u_n, \quad y_n(0) = y_0.$$
It remains to show that $y_n \to y$ weakly in $W^{1,1}$, where y is the solution of
$$y' = a(t, y) + b(t, y) u, \quad y(0) = y = 0.$$
We have
$$|y_n'| \leq |a(t, y_n)| + |b(t, y_n)||u_n| \leq |a(t, 0)| + A(t)|y_n| + |u_n| \big[|b(t, 0)| + B(t)|y_n| \big]$$
where $A(t)$ and $B(t)$ are the Lipschitz constants of $a(t, \cdot)$ and $b(t, \cdot)$. By Gronwall's lemma it follows:
$$|y_n(t)| \leq \left(|y_0| + \int_0^T (|a(t, 0)| + |b(t, 0)||u_n|) \, dt \right) \exp \left(\int_0^T (A(t) + B(t)|u_n|) \, dt \right),$$

which implies that y_n are bounded in L^∞. From the relation $y_n' = a(t, y_n) + b(t, y_n)u_n$ we get that y_n' are equi-uniformly integrable. Therefore, by Dunford-Pettis theorem it follows that y_n' are weakly compact in L^1 and hence y_n are weakly compact in $W^{1,1}$.   ∎

As an application of the results above, consider an optimal control problem governed by an ordinary differential equation (or system), and with integral cost functional, of the form

$$(3.12) \qquad \min\left\{ \int_0^T j(t, y, u)dt \ : \ y' = g(t, y, u), \ y(0) = y_0 \right\}.$$

Here we have taken

- the space Y of states as the space $W^{1,1}(0, T; \mathbb{R}^N)$ of all absolutely continuous functions on $(0, T)$ with values in $\mathbb{R}^N$;

- the space U of controls as the space $L^1(0, T; \mathbb{R}^m)$ of all Lebesgue integrable functions on $(0, T)$ with values in $\mathbb{R}^m$;

- the set $\mathcal{A}$ of admissible pairs as the subset of $U \times Y$ of all pairs (u, y) which satisfy the state equation

$$y' = g(t, y, u) \qquad y(0) = y_0;$$

- the cost functional J as the integral functional

$$J(u, y) = \int_0^T j(t, y, u)\, dt.$$

In order to fulfill the conditions of Lemma 3.5.2, of Proposition 3.5.4, and of Proposition 3.5.5 we make the following assumptions on the data. On the cost integrand j:

A1 the function $j : (0, T) \times \mathbb{R}^N \times \mathbb{R}^m \to [0, +\infty]$ is nonnegative and Borel measurable (or more generally measurable for the σ-algebra $\mathcal{L} \otimes \mathcal{B}_N \otimes \mathcal{B}_m$));

A2 the function $j(t, \cdot, \cdot)$ is lower semicontinuous on $\mathbb{R}^N \times \mathbb{R}^m$ for a. e. $t \in (0, T)$;

A3 the function $j(t, s, \cdot)$ is convex on $\mathbb{R}^N$ for a. e. $t \in (0, T)$ and for every $s \in \mathbb{R}^N$;

A4 there exist $\alpha \in L^1(0, T)$ and $\theta : \mathbb{R} \to \mathbb{R}$, with θ superlinear (that is, $\theta(r)/r \to +\infty$ as $r \to +\infty$) such that

$$\theta(|z|) - \alpha(t) \leq j(t, s, z) \qquad \forall (t, s, z).$$

On the function g in the state equation we assume it is of the form

$$g(t, s, z) = a(t, s) + b(t, s)z,$$

where

A5 the function $a : (0, T) \times \mathbb{R}^N \to \mathbb{R}^N$ is measurable in t and continuous in s, and satisfies

(3.13)
$$\begin{aligned} |a(t, s_2) - a(t, s_1)| &\leq A(t)|s_2 - s_1| \qquad \text{with } A \in L^1(0, T) \\ |a(t, 0)| &\leq M(t) \qquad \text{with } M \in L^1(0, T); \end{aligned}$$

A6 the function $b : (0, T) \times \mathbb{R}^N \to \mathbb{R}^{mN}$ is measurable in t and continuous in s, and satisfies

(3.14)
$$\begin{aligned} |b(t, s_2) - b(t, s_1)| &\leq B|s_2 - s_1| \qquad \text{with } B \in \mathbb{R} \\ |b(t, 0)| &\leq K \text{ with } K \in \mathbb{R}. \end{aligned}$$

The existence result is then the following.

Theorem 3.5.6 *Under assumptions* **A1**–**A6** *above the optimal control problem (3.12) admits at least one solution.*

Proof In order to apply the direct methods of the calculus of variations, we endow the space U of controls with the weak $L^1(0, T; \mathbb{R}^m)$ topology and the space Y of states with the topology of uniform convergence, and we make the following remarks.

• The cost functional J is sequentially lower semicontinuous on $U \times Y$; this follows from the De Giorgi-Ioffe lower semicontinuity theorem for integral functionals. The first proof has been given by De Giorgi in an unpublished paper [102]; another independent proof was given by Ioffe [138]; for a discussion about the lower semicontinuity of integral functionals we refer to the book by Buttazzo [54].

- The functional J is coercive with respect to the variable u; this is a consequence of the Dunford-Pettis weak compactness criterion.

- For every $u \in U$ the state equation

$$y' = a(t, y) + b(t, y)u, \qquad y(0) = y_0$$

has a unique solution $y \in Y$ defined on the whole interval $[0, T]$, thanks to the Lipschitz assumptions made on the coefficients $a(t, \cdot)$ and $b(t, \cdot)$.

- The set $\mathcal{A}$ of admissible pairs is sequentially closed in $U \times Y$ as it can be easily verified by writing the state equation in integral form

$$y(t) = y_0 + \int_0^t \big(a(s, y(s)) + b(s, y(s))u(s)\big) \, ds.$$

By the remarks above, it remains only to prove the coercivity of J on $\mathcal{A}$ with respect to y. In other words, if $u_n \to u$ weakly in $L^1(0, T; \mathbb{R}^m)$ and

$$y_n' = a(t, y_n) + b(t, y_n)u_n , \qquad y_n(0) = y_0,$$

we have to prove that (y_n), or a subsequence of it, converges uniformly. By Gronwall's Lemma we obtain that (y_n) is uniformly bounded, so that by the state equations we obtain

(3.15) $$|y_n'| \leq c(t) + C|u_n|$$

for suitable $c \in L^1(0, T)$ and $C > 0$. Since (u_n) is weakly compact in $L^1(0, T; \mathbb{R}^m)$, by the Dunford-Pettis theorem again, it turns out to be equi-absolutely integrable on $(0, T)$, that is,

$$\forall \varepsilon > 0 \, \exists \delta > 0 \ : \ E \subset (0, T), \ |E| < \delta \ \Rightarrow \ \int_E |u_n| \, dt < \varepsilon \qquad \forall n \in \mathbb{N}.$$

Therefore by (3.15), also (y_n') is equi-absolutely integrable on $(0, T)$, which implies the weak compactness in $L^1(0, T; \mathbb{R}^N)$ of (y_n') and hence the compactness in $L^\infty(0, T; \mathbb{R}^N)$ of (y_n). ∎

When conditions of Theorem 3.5.6 are not fulfilled, we do not have, in general, the existence of a solution of the optimal control problem (3.12), and in order to study the asymptotic behaviour of minimizing sequences (u_n, y_n) we have to consider the associated relaxed formulation.

The simplest case is when we do not have to enlarge the class U of controls, which happens for instance when a coercivity assumption like **A4** is fulfilled. In this case it is enough to take the lower semicontinuous envelope in $U \times Y$ of the mapping

$$(u, y) \mapsto J(u, y) + \chi_A(u, y).$$

In some cases, which often occur in applications to concrete problems, the lower semicontinuous envelope above can be easily computed in terms of the envelope $\overline{J}$ of the cost functional and of the closure $\overline{A}$ of the state equation. More precisely, the following result can be proved.

Proposition 3.5.7 *Assume that*

i) $|J(u, y) - J(u, z)| \leq \omega(y, z)\Phi(u)$ *for every* $u \in U$ *and* $y, z \in Y$ *with* Φ *locally bounded in* U *and*

$$\lim_{z \to y} \omega(y, z) = 0;$$

ii) *if* $(u, y) \in \overline{A}$ *then for every* v *close to* u *there exists* y_v *such that* $(v, y_v) \in A$ *and the mapping* $v \mapsto y_v$ *is continuous.*

Then the relaxed problem associated to

$$\min \{J(u, y) \ : \ (u, y) \in A\}$$

can be written in the form

$$\min \{\overline{J}(u, y) \ : \ (u, y) \in \overline{A}\}.$$

As an example let us consider again an optimal control problem governed by an ordinary differential equation:

(3.16)
$$J(u, y) = \int_0^T j(t, y, u) \, dt$$
$$A = \{(u, y) \in U \times Y \ : \ y' = a(t, y) + b(t, y)\beta(t, u), \ y(0) = y_0\}$$

where the functions a and b satisfy conditions **A5** and **A6**, and β can be nonlinear and j nonconvex with respect to u. If the integrand j is bounded from below by

$$|u|^p - \alpha(t) \leq j(t, y, u) \qquad \text{with } p > 1 \text{ and } \alpha \in L^1,$$

then we may take $U = L^p(0, T; \mathbb{R}^m)$ and $Y = W^{1,1}(0, T; \mathbb{R}^N)$ endowed with their weak topologies. Introducing the auxiliary variable $v = \beta(t, u)$ the new control space is $U \times V$ where V is an L^q space, provided

$$|\beta(t, u)| \le \beta_0(t) + c|u|^{p/q} \qquad \text{with } q > 1 \text{ and } \beta_0 \in L^q,$$

so that the problem can be written in an equivalent form with

(3.17)

$$\tilde{J}(u, v, y) = \int_0^T \left(j(t, y, u) + \chi_{\{v = \beta(t,u)\}} \right) dt$$
$$\tilde{A} = \{(u, v, y) \in U \times V \times Y \ : \ y' = a(t, y) + b(t, y)v, \ y(0) = y_0\}.$$

In this form, we already know that the set $\tilde{A}$ is closed, since the differential equation is now linear in the control. So it remains to relax the cost $\tilde{J}$ with respect to (u, v). If we assume the continuity condition on j:

$$|j(t, y, u) - j(t, z, u)| \le \omega(y, z)(\alpha(t) + |u|^p)$$

is satisfied with $\alpha \in L^1$ and ω such that

$$\lim_{z \to y} \omega(y, z) = 0,$$

then the relaxed form of $\tilde{J}$ is well known and is given by the integral functional

$$\tilde{J}^{**}(u, v, y) = \int_0^T \left(j(t, y, \xi) + \chi_{\{\eta = \beta(t,\xi)\}} \right)^{**}(u, v) \, dt,$$

where the convexification $\tilde{J}^{**}$ is intended with respect to the pair (u, v), and in the integrand with respect to the pair (ξ, η). Finally, eliminating the auxiliary variable v we obtain the relaxed form of the optimal control problem:

$$\min \left\{ \int_0^T \phi(t, y, u, y') \, dt \ : \ u \in L^p(0, T; \mathbb{R}^m), \ y \in W^{1,1}(0, T; \mathbb{R}^N), \ y(0) = y_0 \right\},$$

where the function ϕ takes into account cost and state equation at one time, and is defined by

$$\phi(t, y, u, w) = \inf \left\{ \left(j(t, y, \xi) + \chi_{\{\eta = \beta(t,\xi)\}} \right)^{**}(u, v) \ : \ w = a(t, y) + b(t, y)v \right\}.$$

A case in which the computation can be made explicitely is the following (see Example 5.3.7 of [54]):

(3.18)
$$J(u, y) = \int_0^1 \left(u^2 + \frac{1}{u^2} + |y - y_0|^2 + h(t)u \right) dt$$
$$A = \{(u, y) \in U \times Y : uy' = 1, \; 1/c \le u \le c, \; y(0) \in K\}.$$

Here $y_0(t)$ and $h(t)$ are two functions in $L^2(0, 1)$, $c \ge 1$ is a constant, and K is a closed subset of $\mathbb{R}$. We obtain, after some elementary calculations, that the relaxed problem is the minimization problem for the functional

$$\int_0^1 \left(u^2 + |y'|^2 + 2(uy' - 1) + |y - y_0(t)|^2 + h(t)u \right) dt$$

with the constraints

$$\frac{1}{u} \le y' \le c + \frac{1}{c} - u, \qquad \frac{1}{c} \le u \le c, \qquad y(0) \in K.$$

Consider now the case of a control problem where the control occurs on the coefficient of a second order state equation. More precisely, given $\alpha > 0$ take

(3.19)
$$U = \{u \in L^1(0, 1) : u \ge \alpha \text{ a.e. on } (0, 1)\}$$
$$Y = H_0^1(0, 1) \text{ with the strong topology of } L^2(0, 1)$$

and consider the optimal control problem

(3.20) $\min \left\{ \int_0^1 (g(x, u) + \phi(x, y)) \, dx \; : \; u \in U, \; y \in Y, \; -(uy')' = f \right\}.$

Here $f \in L^2(0, 1)$, and g, ϕ are Borel functions from $(0, 1) \times \mathbb{R}$ into $\overline{\mathbb{R}}$ with

B1 $\phi(x, \cdot)$ is continuous on $\mathbb{R}$ for a.e. $x \in (0, 1)$,

B2 for a suitable function $w(x, t)$ integrable in x and increasing in t we have

$$|\phi(x, s)| \le w(x, |s|) \qquad \forall (x, s) \in (0, 1) \times \mathbb{R}.$$

Setting for any $(u, y) \in U \times Y$

(3.21)
$$J(u, y) = \int_0^1 (g(x, u) + \phi(x, y)) \, dx$$
$$G(u, y) = \int_0^1 (uy'^2 - 2fy) \, dx$$

we obtain that problem (3.20) can be written in the form

$$\min \left\{ J(u,y) \; : \; u \in U, \; y \in Y, \; y \in \operatorname{argmin} G(u, \cdot) \right\}.$$

It is well-known that

$$\Gamma \lim_{n \to +\infty} G(u_n, \cdot) = G(u, \cdot) \qquad \Longleftrightarrow \qquad \frac{1}{u_n} \to \frac{1}{u} \text{ weakly* in } L^\infty(0,1);$$

therefore, by applying the framework of Section 3.5 we obtain $\hat{U} = U$, $\hat{G} = G$, and

$$\hat{J}(u,y) = \int_0^1 \left(\gamma(x,u) + \phi(x,y) \right) dx$$

where $\gamma(x,s) = \beta^{**}(x, 1/s)$ being ** the convexification operator (with respect to the second variable) and

$$(3.22) \qquad\qquad \beta(x,t) = \begin{cases} g(x, 1/t) & if\, t \in]0, 1/\alpha] \\ +\infty & otherwise. \end{cases}$$

For instance, if $\alpha < 1$ and $g(x,s) = |s - 1|$ we have

$$(3.23) \qquad\qquad \beta(x,t) = \begin{cases} s - 1 & if s \geq 1 \\ \alpha(1 - s)/s & if \alpha \leq s < 1. \end{cases}$$

An analogous computation can be done in the case

$$U = \left\{ u \in L^1(0,1) \; : \; u \geq 0, \; \int_0^1 \frac{1}{u}\, dx \leq c \right\},$$

where $c > 0$. In this case, in order to satisfy the coercivity assumption required by the abstract framework, it is better to consider

$Y = BV(0,1)$ with the strong topology of $L^1(0,1)$,
$G(u,y) = \int_0^1 \left(uy'^2 - 2fy \right) dx + \chi_{\{y(0)=y_0, \; y(1)=y_1\}}(y) + \chi_{\{y' << dx\}}(y)$

where $y' << dx$ denotes the constraint that y' is a measure absolutely continuous with respect to the Lebesgue measure. Following Buttazzo and Freddi

[5] we obtain that $\hat{U}$ coincides with the set of positive measures μ on $[0, 1]$ such that $\mu([0, 1]) \le c$ and

$$
\hat{G}(\mu, y) = \int_{]0,1[} \left| \frac{dy'}{d\mu} \right|^2 d\mu - 2 \int_{]0,1[} fy \, dx
$$
$$
+ \frac{|y^+(0) - y_0|^2}{\mu(\{0\})} + \frac{|y^+(1) - y_1|^2}{\mu(\{1\})} + \chi_{\{y' << \mu\}}(y)
$$

where $dy'/d\mu$ is the Radon-Nikodym derivative of y' with respect to μ. It is not difficult to see that all assumptions required by the abstract framework are fulfilled, with

$$
\psi(y) = a\|y\|_{BV}^2 - b
$$

for suitable positive constants a, b. It remains to compute the functional $\hat{J}$. Assume for simplicity that $g(x, s) = g(s)$ and that **B1** and **B2** hold; then we obtain

$$
\hat{J}(\mu, y) = \int_0^1 \beta^{**}(\mu^a(x)) \, dx + \int_{[0,1]} (\beta^{**})^\infty(\mu^s) + \int_0^1 \phi(x, y) \, dx
$$

where $\beta(t) = g(1/t)$, $\mu = \mu^a \cdot dx + \mu^s$ is the Lebesgue-Nikodym decomposition of μ, and $(\beta^{**})^\infty$ is the recession function of β^{**}. For instance, if $g(s) = |s - 1|^2$, the relaxed problem has the form

$$
\min \left\{ \int_{\{\mu^a \le 1\}} \left| \frac{\mu^a - 1}{\mu^a} \right|^2 dx + \int_0^1 \phi(x, y) \, dx \; : \; \mu([0, 1]) \le c, \; y \in \operatorname{argmin} \hat{G}(\mu, \cdot) \right\}.
$$

3.6 Examples of relaxed shape optimization problems

In this section we give some applications of the abstract framework for relaxed controls introduced in the previous sections, and we characterize explicitly the set $\hat{U}$ of relaxed controls as well as the form of the relaxed control problems.

Example 3.6.1 The first example deals with a class of shape optimization problems with Dirichlet conditions on the free boundary, studied by Buttazzo and Dal Maso (see [62], [63], [64], [65] and references therein), in which the initial set U of controls is the class of all domains contained in a given open subset D of $\mathbb{R}^N$. We stress that this class has no linear or convex structure, and the usual topologies on families of domains are not suitable for the problems one would like to consider.

To set the problem more precisely, let D be a bounded open subset of $\mathbb{R}^N$ ($N \geq 2$), let $f \in L^2(D)$ and let $j : D \times \mathbb{R} \to \mathbb{R}$ be a Borel function. Consider the shape optimization problem

$$(3.24) \qquad \min\left\{ \int_D j(x, y_A(x))\, dx \ : \ A \in \mathcal{A}(D)\right\}$$

where $\mathcal{A}(D)$ is the family of all open subsets of D and where for every $A \in \mathcal{A}(D)$ we denoted by y_A the solution of the Dirichlet problem

$$(3.25) \qquad \begin{cases} -\Delta y = f \text{ in } A \\ y \in H_0^1(A) \end{cases}$$

extended by zero to $D \setminus A$. In this way all states y belong to the Sobolev space $H_0^1(D)$, which will be taken as the space of states. The setting of the optimal control problem we consider is then:

- the space of states is $Y = H_0^1(D)$ with the strong topology of $L^2(D)$;

- the set of controls is $U = \mathcal{A}(D)$;

- the cost functional is taken of the integral form

$$J(A, y) = \int_D j(x, y(x))\, dx \ ;$$

notice that in this case the cost does not explicitly depend on the control variable A;

- the state functional is

$$(3.26) \qquad G(A, y) = \int_D \left(|\nabla y|^2 - 2fy\right) dx + \chi_{H_0^1(A)}(y)$$

which provides, via Euler-Lagrange equation, the state equation (3.25). The shape optimization problem with Dirichlet conditions on the free boundary can then be written as an optimal control problem, in the form

$$(3.27) \qquad \min\left\{J(y) \ : \ y \in Y, \ A \in U, \ y \in \operatorname{argmin} G(A, \cdot)\right\}.$$

It is easy to verify that, as a consequence of Poincaré inequality, the coercivity assumption required in Definition 3.3.1 ii) turns out to be fulfilled, with

$$\Psi(y) = C_1 \int_D |\nabla y|^2 dx - C_2$$

for suitable positive constants C_1 and C_2.

In order to identify the relaxed problem associated with (3.27) we have first to characterize the completion $\hat{U}$ of U with respect to the distance induced by the Γ-convergence on the functionals $G(A, \cdot)$. This has been done by Dal Maso and Mosco in [97], where it is shown that $\hat{U}$ coincides with the space $\mathcal{M}_0(D)$ of all nonnegative Borel measures, possibly $+\infty$ valued, which vanish on all sets of capacity zero. The identification of the relaxed state functional $\hat{G}$ has also been given and, for every $\mu \in \mathcal{M}_0(D)$ and $y \in H^1_0(D)$ we have

$$\hat{G}(\mu, y) = \int_D \left(|\nabla y|^2 - 2fy \right) dx + \int_D y^2 \, d\mu.$$

The relation $y \in \operatorname{argmin} \hat{G}(\mu, \cdot)$ can also be written, via Euler-Lagrange equation, in the form

(3.28)
$$\begin{cases} -\Delta y + \mu y = f \text{ in } D \\ \quad\quad y \in H^1_0(D) \end{cases}$$

which has to be intended in the following weak sense: $y \in H^1_0(D) \cap L^2(D, \mu)$ and

$$\int_D \nabla y \nabla \varphi \, dx + \int_D y\varphi \, d\mu = \int_D f\varphi \, dx \quad \forall \varphi \in H^1_0(D) \cap L^2(D, \mu).$$

On the integrand j appearing in the cost functional J we make the following assumptions:

i) $j(x, \cdot)$ is continuous for a.e. $x \in D$;

ii) for suitable $a \in L^1(D)$ and $b \in \mathbb{R}$ we have $|j(x, s)| \le a(x) + b|s|^2$ for a.e. $x \in D$ and for every $s \in \mathbb{R}$.

In this way the functional J turns out to be continuous in the strong topology of L^2, so that the assumptions of the abstract scheme apply, and the corresponding relaxed problem can be written in the form

$$\min \left\{ \int_D j(x, y(x)) \, dx \; : \; \mu \in \mathcal{M}_0(D), \; y \in H^1_0(D), \; -\Delta y + \mu y = f \right\}.$$

Example 3.6.2 The second example we consider is the case of a control problem where the control occurs on the coefficient of the state equation, which is of partial differential type. More precisely, we consider a class of optimal control problems for two-phase conductors which has been studied by Cabib and Dal Maso (see [74], [75]). As in the case seen in Section 3.5, also here the control occurs on the coefficients of the state equation which is actually an elliptic partial differential equation. More precisely, let D be a bounded open subset of $\mathbb{R}^n$, let α, β be two real positive numbers, $f \in L^2(D)$. We consider the class U of controls as the set of all functions $u : D \to \mathbb{R}$ with the property that there exists a Borel subset $A \subset D$ such that

$$u = \alpha 1_A + \beta 1_{D \setminus A}.$$

In this way we can identify the class U with the family of all Borel subsets of D. The space of states will be $Y = H_0^1(D)$ endowed with the strong topology of L^2.

Consider the optimal control problem

$$\min \left\{ J(u,y) \ : \ -\operatorname{div}(u \nabla y) = f \text{ in } D, \ u = 0 \text{ on } \partial D \right\},$$

where the cost functional is still of the form

$$J(u,y) = \int_A g(x)\, dx + \int_D \varphi(x,y)\, dx,$$

where g is a given function in $L^1(D)$, and $\varphi : D \times \mathbb{R} \to \mathbb{R}$ is a Charathéodory integrand which satisfies the growth condition

$$|\varphi(x,z)| \le c_1(x) + c_2 z^2 \qquad \text{for suitable } c_1 \in L^1(D), \ c_2 \ge 0.$$

The energy functional G is now given by

$$G(u,y) = \int_D \left(u |\nabla y|^2 - 2fy \right) dx.$$

The completion $\hat{U}$ of U with respect to the G-convergence of the state equation or, equivalently, the Γ-convergence of the functionals $G(u, \cdot)$ has been characterized by Lurie and Cherkaev [152], [153] for the two-dimensional case and by Murat and Tartar [164], [178] for the general case. They proved that

$\hat{U}$ is the space of all symmetric $n \times n$ matrices $A(x) = (a_{ij}(x))$ whose eigenvalues $\lambda_1(x) \leq \lambda_2(x) \leq \cdots \leq \lambda_n(x)$ satisfy for a suitable $t \in [0,1]$ (depending on x) the following $n + 2$ inequalities:

$$\sum_{i=1}^{n} \frac{1}{\lambda_i - \alpha} \leq \frac{1}{\nu_t - \alpha} + \frac{n-1}{\mu_t - \alpha}$$

$$\sum_{i=1}^{n} \frac{1}{\beta - \lambda_i} \leq \frac{1}{\beta - \nu_t} + \frac{n-1}{\beta - \mu_t}$$

$$n\mu_t \leq \lambda_i \leq \mu_t, \qquad i = 1, \ldots, n,$$

where μ_t and ν_t respectively denote the arithmetic and the harmonic mean of α and β, namely

$$\mu_t = t\alpha + (1-t)\beta$$
$$\nu_t = \left(\frac{t}{\alpha} + \frac{1-t}{\beta}\right)^{-1}.$$

For instance, when $n = 2$, then $\hat{U}$ consists of all symmetric 2×2 matrices $a(x)$ whose eigenvalues $\lambda_1(x)$, $\lambda_2(x)$ belong, for every $x \in D$, to the following convex domain C of $\mathbb{R}^2$:

$$C = \left\{(\lambda_1, \lambda_2) \in [\alpha, \beta] \times [\alpha, \beta] : \frac{\alpha\beta}{\beta + \alpha - \lambda_1} \leq \lambda_2 \leq \alpha + \beta - \frac{\alpha\beta}{\lambda_1}\right\}.$$

The picture of the set C in dimension 2 is below, with $\alpha = 1$ and $\beta = 2$.

The functional $\hat{G}$ can be computed and we have

$$\hat{G}(A, y) = \int_D \left[a(x)DyDy - 2fy\right] dx.$$

The computation of the functional $\hat{J}$ can be found in Cabib [74] where it is shown that

$$\hat{J}(A, y) = \int_D \left[\hat{g}(x, A) + \varphi(x, y)\right] dx,$$

with

(3.29)
$$\hat{g}(x, A) = \begin{cases} g(x)\dfrac{\beta - \mu(A)}{\beta - \alpha} & \text{if } g(x) \leq 0 \\[2mm] g(x)\dfrac{\beta - \overline{\mu}(A)}{\beta - \alpha} & \text{if } g(x) \geq 0 \end{cases}$$

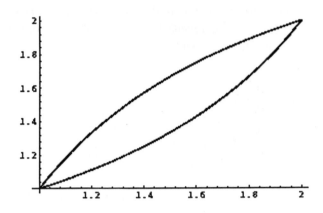

Figure 3.1: The set C in dimension 2.

being

$$\underline{\mu}(A) = \max\left\{\lambda_n,\ \beta + \frac{(n-1)\beta + \alpha}{1 - \beta \sum_{i=1}^{n} \left(\beta - \lambda_i\right)^{-1}}\right\}$$

$$\overline{\mu}(A) = \alpha + \frac{(n-1)\alpha + \beta}{1 + \alpha \sum_{i=1}^{n} \left(\lambda_i - \alpha\right)^{-1}}\ .$$

Chapter 4

Shape optimization problems with Dirichlet condition on the free boundary

4.1 A short survey on capacities

Throughout the next chapters we shall often use the notion of Sobolev capacity defined by

$$\mathrm{Cap}(E) = \inf\Big\{\int_{\mathbb{R}^N} |\nabla u|^2 + u^2\, dx\ :\ u \in \mathcal{U}_E\Big\},$$

where $\mathcal{U}_E$ is the set of all functions u of the Sobolev space $H^1(\mathbb{R}^N)$ such that $u \geq 1$ almost everywhere in a neighborhood of E.

Sometimes it is more convenient to work with a local capacity. Let D be a bounded open set. The capacity of a subset E in D is

$$\mathrm{cap}(E, D) = \inf\Big\{\int_D |\nabla u|^2\, dx\ :\ u \in \mathcal{U}_E\Big\},$$

where $\mathcal{U}_E$ is the set of all functions u of the Sobolev space $H_0^1(D)$ such that $u \geq 1$ almost everywhere in a neighborhood of E. When there is no ambiguity in the choice of D, we simply write $\mathrm{cap}(E)$ instead of $\mathrm{cap}(E, D)$. Since the two capacities defined are "locally equivalent", there is no important difference for our purposes. Nevertheless, in order to avoid any ambiguity, in all definitions given below we consider the first one.

If a property $P(x)$ holds for all $x \in E$ except for the elements of a set $Z \subset E$ with $\text{Cap}(Z) = 0$, we say that $P(x)$ holds *quasi-everywhere* on E (shortly *q.e.* on E). The expression *almost everywhere* (shortly *a.e.*) refers, as usual, to the Lebesgue measure.

A subset A of $\mathbb{R}^N$ is said to be *quasi-open* (resp. *quasi-closed*) if for every $\varepsilon > 0$ there exists an open (resp. closed) subset A_ε of $\mathbb{R}^n$, such that $\text{Cap}(A_\varepsilon \triangle A) < \varepsilon$, where $\triangle$ denotes the symmetric difference of sets. The class of all quasi-open subsets of D will be denoted by $\mathcal{A}$.

A function $f \colon D \to \mathbb{R}$ is said to be *quasi-continuous* (resp. *quasi-lower semicontinuous*) if for every $\varepsilon > 0$ there exists a continuous (resp. lower semicontinuous) function $f_\varepsilon \colon D \to \mathbb{R}$ such that $\text{Cap}(\{f \neq f_\varepsilon\}) < \varepsilon$, where $\{f \neq f_\varepsilon\} = \{x \in D : f(x) \neq f_\varepsilon(x)\}$. It is well known (see, e.g., Ziemer [185]) that every function u of the Sobolev space $H^1(D)$ has a quasi-continuous representative, which is uniquely defined up to a set of capacity zero. We shall always identify the function u with its quasi-continuous representative, so that a pointwise condition can be imposed on $u(x)$ for quasi-every $x \in D$. Notice that with this convention we have

$$\text{cap}(E, D) = \min\left\{ \int_D |\nabla u|^2 dx : u \in H_0^1(D),\ u \geq 1 \text{ q.e. on } E \right\}$$

for every subset E of D.

We recall the following theorem from [3].

Theorem 4.1.1 *Let $u \in H^1(\mathbb{R}^N)$. Then for q.e. $x \in \mathbb{R}^N$*

$$\lim_{\varepsilon \to 0} \frac{\int_{B_{x,\varepsilon}} u(y) dy}{|B_{x,\varepsilon}|} = \tilde{u}(x),$$

where $\tilde{u}$ is a quasi-continuous representative of u.

For every $A \in \mathcal{A}$ we denote by $H_0^1(A)$ the space of all functions $u \in H_0^1(D)$ such that $u = 0$ q.e. on $D \setminus A$, with the Hilbert space structure inherited from $H_0^1(D)$. Note that $H_0^1(A)$ is a closed subspace of $H_0^1(D)$ as a consequence of well known properties of quasi-continuous representatives of Sobolev functions (see, e.g., Ziemer [185]). If A is open, the previous definition of $H_0^1(A)$ is equivalent to the usual one (see Adams-Hedberg [3]). Indeed, we recall the following result.

Theorem 4.1.2 *Let $A \subseteq \mathbb{R}^N$ be an open set. Then a function $u \in H^1(\mathbb{R}^N)$ belongs to $H_0^1(A)$ if and only if $u = 0$ q.e. on A^c (in the previous equality u is supposed to be quasi-continuous).*

We also recall that from every strongly convergent sequence in $H^1(\mathbb{R}^N)$ one can extract a subsequence converging q.e. in $\mathbb{R}^N$. Moreover, we have (see [38]).

Lemma 4.1.3 *If C_1, C_2 are two quasi-open sets with $\mathrm{Cap}(C_1 \cap C_2) = 0$ and $u \in H^1_0(C_1 \cup C_2)$, then $u_{|C_1} \in H^1_0(C_1)$ and $u_{|C_2} \in H^1_0(C_2)$.*

The fine topology on B is the coarsest topology making all super-harmonic functions continuous. The relation between the quasi-topology and the fine topology is studied in [121], [140], [3]. We recall the following theorem from [140].

Theorem 4.1.4 *Suppose $A \subseteq \mathbb{R}^N$. Then the following assertions are equivalent:*

i) A is quasi-open;

ii) A is the union of a finely open set and a set of zero capacity;

iii) $A = \{u > 0\}$ for some nonnegative quasi-continuous function $u \in H^1(\mathbb{R}^N)$.

Since the family of quasi-open sets of $\mathbb{R}^N$ is not a topology (only countable unions of quasi-open sets are quasi-open) when dealing with arbitrary unions of quasi-open sets, sometimes it is more interesting to work with the finely open sets given by the previous theorem at point ii).

To finish, we also recall the following:

Theorem 4.1.5 *Suppose A is a quasi-open subset of $\mathbb{R}^N$ and u is a function on A. The following assertions are equivalent:*

i) u is quasi-l.s.c.;

ii) the sets $\{u > c\}$ are quasi-open for all $c \in \mathbb{R}$;

iii) u is finely l.s.c. up to a set of zero capacity.

In order to discuss the relaxation of Dirichlet problems we denote by $\mathcal{M}_0(D)$ the set of all non-negative Borel measures μ on D, possibly $+\infty$ valued, such that

i) $\mu(B) = 0$ for every Borel set $B \subseteq D$ with $\mathrm{Cap}(B) = 0$

ii) $\mu(B) = \inf\{\mu(U) : U \text{ quasi-open}, B \subseteq U\}$ for every Borel set $B \subseteq D$.

4.2   Nonexistence of optimal solutions

In this section we give an explicit example where the existence of an optimal domain does not occur (see also chapter 3). The shape optimization problem we consider is with Dirichlet conditions on the free boundary, of the form

$$(4.1) \qquad \min\left\{ J(u_A) \ : \ -\Delta u_A = f \text{ in } A, \ u_A \in H_0^1(A) \right\}.$$

Here the admissible domains A vary in the class of all subdomains of a given bounded open subset D of $\mathbb{R}^N$, $f \in L^2(D)$ is fixed, and the solutions u_A are considered extended by zero on $D \setminus A$.

The cost functional we consider is the $L^2(D)$ distance from a desired state $\bar{u}(x)$

$$(4.2) \qquad\qquad J(u) = \int_D |u - \bar{u}|^2 \, dx.$$

In the thermostatic model the optimization problem (4.1) consists in finding an optimal distribution, inside D, of the Dirichlet region $D \setminus A$ in order to achieve a temperature which is as close as possible to the desired temperature $\bar{u}$, once the heat sources f are prescribed.

For simplicity, we consider an uniformly distributed heat source, that is we take $f \equiv 1$, and we take the desired temperature $\bar{u}$ constantly equal to $c > 0$. Therefore problem (4.1) becomes

$$(4.3) \qquad \min\left\{ \int_D |u_A - c|^2 \, dx \ : \ -\Delta u_A = 1 \text{ in } A, \ u_A \in H_0^1(A) \right\}.$$

We will actually prove that for small values of the constant c no regular domain A can solve problem (4.3) above; the proof of nonexistence of any domain is slightly more delicate and requires additional tools like the capacitary form of necessary conditions of optimality (see for instance [62], [63], [82]).

Proposition 4.2.1 *If c is small enough, then no regular domain A can solve the optimization problem (4.3).*

Proof   Assume by contradiction that a regular domain A solves the optimization problem (4.3). Let us also assume first that A does not coincide with the whole set D, so that we can take a point x_0 in D which does not belong to the closure $\overline{A}$ and a small ball B_ε of radius ε, centered at x_0 and disjoint form A. If u_A denotes the solution of

(4.4)
$$\begin{cases} -\Delta u = 1 & \text{in } A \\ u \in H_0^1(A) \end{cases}$$

then the solution $u_{A\cup B_\varepsilon}$, corresponding to the admissible choice $A \cup B_\varepsilon$, can be easily identified, and we find

(4.5)
$$u_{A\cup B_\varepsilon}(x) = \begin{cases} u_A(x) & \text{if } x \in A \\ (\varepsilon^2 - |x - x_0|^2)/4 & \text{if } x \in B_\varepsilon \\ 0 & \text{otherwise.} \end{cases}$$

Therefore, we obtain

$$J(u_A) = \int_A |u_A - c|^2\, dx + \int_{B_\varepsilon} c^2\, dx + \int_{D\backslash(A\cup B_\varepsilon)} c^2\, dx$$

$$J(u_{A\cup B_\varepsilon}) = \int_A |u_A - c|^2\, dx + \int_{B_\varepsilon} \left| \frac{\varepsilon^2 - |x - x_0|^2}{4} - c \right|^2 dx + \int_{D\backslash(A\cup B_\varepsilon)} c^2\, dx.$$

Comparing the cost $J(u_A)$ with the cost $J(u_{A\cup B_\varepsilon})$ and using the minimality of A then gives

$$c^2 \operatorname{meas}(B_\varepsilon) \leq \int_{B_\varepsilon} \left| \frac{\varepsilon^2 - |x - x_0|^2}{4} - c \right|^2 dx$$

$$= N\varepsilon^{-N} \operatorname{meas}(B_\varepsilon) \int_0^\varepsilon \left| \frac{\varepsilon^2 - r^2}{4} - c \right|^2 r^{N-1}\, dr$$

$$= c^2 \operatorname{meas}(B_\varepsilon) + \frac{1}{16} \int_0^\varepsilon (\varepsilon^2 - r^2)(\varepsilon^2 - r^2 - 8c)r^{N-1}\, dr$$

which, for a fixed $c > 0$, turns out to be false if ε is small enough.

Thus all regular domains $A \neq D$ are ruled out by the argument above. We can now exclude also the case $A = D$ if c is small, by comparing for instance the full domain D with the empty set. This gives, taking into account that $u_\emptyset \equiv 0$,

(4.6)
$$J(u_D) = \int_D |u_D - c|^2\, dx$$
$$J(u_\emptyset) = \int_D c^2\, dx$$

so that we have $J(u_\emptyset) < J(u_D)$ if c is sufficiently small. Hence all regular subdomains of D are excluded, and the proof is complete.   ∎

Example 4.2.2 If we take into account the identification of the class of relaxed domains seen in Section 3.6, then we may produce, rather simply, other examples of nonexistence of optimal domains. Take indeed a smooth function f in (4.1) such that $f(x) > 0$ in D and let w be the solution of the problem

$$(4.7) \qquad \begin{cases} -\Delta w = f \text{ in } D \\ \quad w \in H_0^1(D). \end{cases}$$

It is well known, from the maximum principle, that $w(x) > 0$ in D. Take now the desired state $\bar{u}(x) = w(x)/2$ and the cost density $j(x, s) = |s - \bar{u}(x)|^2$ like in (4.2). Then the optimization problem

$$\min\left\{ \int_D |u_A - \bar{u}|^2 \, dx \; : \; -\Delta u_A = f \text{ in } A, \; u_A \in H_0^1(A) \right\}$$

admits the relaxed formulation

$$\min\left\{ \int_D |u_\mu - \bar{u}|^2 \, dx \; : \; -\Delta u_\mu + u\mu = f \text{ in } D, \; u_\mu \in H_0^1(D) \right\}$$

where the measure μ varies now in the class of relaxed controls seen in Section 3.6. It is easy to see that the relaxed problem attains its minimum value 0 at the measure

$$\mu = (f/w) \cdot dx$$

which corresponds to the solution $u_\mu = w/2$ of the relaxed state equation

$$(4.8) \qquad \begin{cases} -\Delta u_\mu + u\mu = f \text{ in } D \\ \quad u_\mu \in H_0^1(D). \end{cases}$$

On the other hand, since $\bar{u} > 0$ in D, it is clear that there are no domains $A \neq D$ such that $u_A = \bar{u}$ in D. The case $A = D$ has also to be excluded, because $u_D = w > w/2 = \bar{u}$.

The assumption above that f is smooth can be weakened by simply requiring that $f(x) > 0$ for a.e. $x \in D$.

4.3 The relaxed form of a Dirichlet problem

As already seen in Section 3.6 the relaxed form of a shape optimization problem with Dirichlet conditions on the free boundary involves relaxed controls which are measures. In this section we give more details about this topic; the reader may find a complete discussion in [63].

We know that the definition of relaxed controls only depends on the state equation, that we take for simplicity of the form

$$-\Delta u = f \text{ in } A, \qquad u \in H_0^1(A).$$

Here the control variable A runs in the class of open subsets of a given bounded subset D of $\mathbb{R}^N$ and f is a given function in $L^2(D)$. As already stated in Section 3.6 the class of relaxed controls is the class $\mathcal{M}_0(D)$ of all nonnegative Borel measures μ on D such that $\mu(B) = 0$ for every Borel subset B of D such that $\text{cap}(B) = 0$. We stress the fact that the measures $\mu \in \mathcal{M}_0(D)$ do not need to be finite, and may take the value $+\infty$ even on large parts of D.

For every measure $\mu \in \mathcal{M}_0(D)$ we denote by A_μ the *"set of finiteness"* of μ; more precisely A_μ is defined as the union of all finely open subsets A of D such that $\mu(A) < +\infty$; A_μ is called the regular set of the measure μ. By its definition the set A_μ is finely open, hence quasi-open. We also denote by $S_\mu = D \setminus A_\mu$ the singular set of μ.

For example, if $N - 2 < \alpha \le N$ the α-dimensional Hausdorff measure $\mathcal{H}^\alpha$ belongs to $\mathcal{M}_0(D)$; in fact every Borel set with capacity zero has an Hausdorff dimension which is less than or equal to $N - 2$. Another example of measure of the class $\mathcal{M}_0(D)$ is, for every $S \subset D$, the measure ∞_S defined by

$$(4.9) \qquad \infty_S(B) = \begin{cases} 0 & \text{if } \text{cap}(B \cap S) = 0, \\ +\infty & \text{otherwise.} \end{cases}$$

In order to write correctly the relaxed form of the state equation we introduce the space $X_\mu(D)$ as the vector space of all functions $u \in H_0^1(D)$ such that $\int_D u^2 \, d\mu < \infty$. Note that, since μ vanishes on all sets with capacity zero and since Sobolev functions are defined up to sets of capacity zero, the definition of $X_\mu(D)$ is well posed. In other words we may think to $X_\mu(D)$ as to $H_0^1(D) \cap L^2(D, \mu)$; moreover we can endow $X_\mu(D)$ with the norm

$$\|u\|_{X_\mu(D)} = \left(\int_D |\nabla u|^2 \, dx + \int_D u^2 \, d\mu \right)^{1/2}$$

which comes from the scalar product

$$(u, v)_{X_\mu(D)} = \int_D \nabla u \nabla v \, dx + \int_D uv \, d\mu.$$

It is possible to show (see [63]) that with the scalar product above the space $X_\mu(D)$ becomes a Hilbert space.

Since $X_\mu(D)$ can be embedded into $H_0^1(D)$ by the identity mapping $i(u) = u$, the dual space $H^{-1}(D)$ of $H_0^1(D)$ can be considered as a subspace of the dual space $X_\mu'(D)$. We then write for $f \in H^{-1}(D)$

$$\langle f, v \rangle_{X_\mu'(D)} = \langle f, v \rangle_{H^{-1}(D)} \qquad \forall v \in X_\mu(D)$$

and so, when $f \in L^2(D)$

$$\langle f, v \rangle_{X_\mu'(D)} = \int_D fv \, dx \qquad \forall v \in X_\mu(D).$$

Example 4.3.1 Take $\mu = a(x)\mathcal{H}^N$ where $a \in L^p(D)$ and

$$(4.10) \qquad \begin{cases} N/2 \leq p \leq +\infty & \text{if } N \geq 3 \\ 1 < p \leq +\infty & \text{if } N = 2. \end{cases}$$

Then, by the Sobolev embedding theorem and Hölder inequality, we have that $X_\mu(D) = H_0^1(D)$ with equivalent norms.

Example 4.3.2 Let A be a finely open subset of D and let $S = D \setminus A$; take $\mu = \infty_S$ as defined in (4.9). Then, by the Poincaré inequality, we have that $X_\mu(D) = H_0^1(D)$ with equivalent norms. The same conclusion holds if $\mu = \infty_S + a(x)\mathcal{H}^N$ where $a \in L^p(D)$ with p satisfying the conditions of the previous example.

Consider now a measure $\mu \in \mathcal{M}_0(D)$. By the Riesz representation theorem, for every $f \in X_\mu'(D)$ there exists a unique $u \in X_\mu(D)$ such that

$$(4.11) \qquad (u, v)_{X_\mu(D)} = \langle f, v \rangle_{X_\mu'(D)} \qquad \forall v \in X_\mu(D).$$

By the definition of scalar product in $X_\mu(D)$ this turns out to be equivalent to

$$(4.12) \qquad \int_D \nabla u \nabla v \, dx + \int_D uv \, d\mu = \langle f, v \rangle_{X_\mu'(D)} \qquad \forall v \in X_\mu(D)$$

that we simply write in the form

$$u \in X_\mu(D), \qquad -\Delta u + \mu u = f \text{ in } X'_\mu(D).$$

This is the relaxed state equation of the optimal control problem we shall consider. In other words, the *resolvent operator* $R_\mu : X'_\mu(D) \to X_\mu(D)$ which associates to every $f \in X'_\mu(D)$ the unique solution u of (4.12) is well defined. Moreover it easy to see that it is linear and continuous from $X'_\mu(D)$ onto $X_\mu(D)$, it is symmetric, that is

$$\langle g, R_\mu(f) \rangle_{X'_\mu(D)} = \langle f, R_\mu(g) \rangle_{X'_\mu(D)} \qquad \forall f, g \in X'_\mu(D),$$

and there exists a constant c, which depends only on D, such that

$$\|R_\mu(f)\|_{H^1(D)} \leq c\|f\|_{H^{-1}(D)} \qquad \forall f \in H^{-1}(D).$$

Example 4.3.3 If we take $\mu = a(x)\mathcal{H}^N$ with $a \in L^p$ and p satisfying the assumption of Example 4.3.1, and $f \in H^{-1}(D)$, then, according to what seen in Example 4.3.1, the relaxed state equation simply becomes

$$u \in H^1_0(D), \qquad -\Delta u + au = f \text{ in } H^{-1}(D).$$

Notice that in this case we have $au \in H^{-1}(D)$.

Example 4.3.4 If we take $\mu = \infty_{D \setminus A}$ with A open subset of D, and $f \in H^{-1}(D)$, then, according to what seen in Example 4.3.2, the relaxed state equation simply becomes

$$u \in H^1_0(A), \qquad -\Delta u = f \llcorner A \text{ in } H^{-1}(A)$$

where the restriction $f \llcorner A$ is defined by

$$\langle f \llcorner A, v \rangle_{H^{-1}(A)} = \langle f, v \rangle_{H^{-1}(D)} \qquad \forall v \in H^1_0(A).$$

Example 4.3.5 If we take $\mu = \infty_{D \setminus A} + a(x)\mathcal{H}^N$ with a and A as in the examples above, then the relaxed state equation takes the form

$$u \in H^1_0(A), \qquad -\Delta u + au = f \llcorner A \text{ in } H^{-1}(A).$$

In Section 3.6 we have already stated the fact that the class $\mathcal{M}_0(D)$ is the class of relaxed controls obtained through the abstract relaxation procedure introduced in Section 3.5. In particular, $\mathcal{M}_0(D)$ can be endowed with the topology of γ-convergence, which can be also defined through the resolvent operators.

Definition 4.3.6 *We say that a sequence (μ_n) of measures in $\mathcal{M}_0(D)$ γ-converges to a measure $\mu \in \mathcal{M}_0(D)$ if and only if*

$$R_{\mu_n}(f) \to R_\mu(f) \text{ weakly in } H^1_0(D) \qquad \forall f \in H^{-1}(D).$$

The following properties follow from the abstract scheme introduced in Section 3.5.

Proposition 4.3.7 *The space $\mathcal{M}_0(D)$, endowed with the topology of γ-convergence, is a compact metric space. Moreover, the class of measures of the form $\infty_{D\setminus A}$, with A open (and smooth) subset of D, is dense in $\mathcal{M}_0(D)$.*

Remark 4.3.8 It is easy to see that also the class of measures of the form $a(x)\mathcal{H}^N$, where a is a nonnegative and smooth function in D, is dense in $\mathcal{M}_0(D)$.

Example 4.3.9 An explicit constructive way to approximate every measure μ of $\mathcal{M}_0(D)$ by a sequence of measures of the form $\infty_{D\setminus A_n}$ is given in [96].

In the sequel we show (without proofs) how the relaxed form can be found in a direct way. For this approach we refer to [100]. We also refer the reader to the classical example of Cioranescu and Murat [83] which is briefly presented below.

Let $f \in L^2(D)$ and let (A_n) be a sequence of quasi-open subsets of the bounded design region D. We denote by u_n the solution of the following equation on A_n:

(4.13)
$$\begin{cases} -\Delta u_n = f \text{ in } A_n \\ \quad u_n \in H^1_0(A_n). \end{cases}$$

Suppose that w_n is the solution on A_n of the same equation, but for the right hand side $f \equiv 1$. Extracting a subsequence if necessary, we may suppose that $u_n \rightharpoonup u$ and $w_n \rightharpoonup w$ weakly in $H^1_0(D)$.

Let $\varphi \in C^\infty_0(D)$. Taking as a test function $w_n\varphi$ for (4.13) on A_n we have the following sequence of equalities:

$$\int_D \nabla u_n \nabla (w_n \varphi) dx = \int_D f w_n \varphi dx,$$

$$\int_D \nabla u_n \nabla \varphi w_n dx + \int_D \nabla u_n \nabla w_n \varphi dx = \int_D f w_n \varphi dx,$$

$$\int_D \nabla u_n \nabla \varphi w_n dx - \int_D u_n \nabla w_n \nabla \varphi dx - < \Delta w_n, \varphi u_n >_{H^{-1}(D) \times H_0^1(D)}$$

$$= \int_D f w_n \varphi dx,$$

$$\int_D \nabla u_n \nabla \varphi w_n dx - \int_D u_n \nabla w_n \nabla \varphi dx + \int_D u_n \varphi dx = \int_D f w_n \varphi dx.$$

Making $n \to \infty$ we get

$$\int_D \nabla u \nabla \varphi w dx - \int_D u \nabla w \nabla \varphi dx + \int_D u \varphi dx = \int_D f \varphi w dx.$$

Since

$$- \int_D u \nabla w \nabla \varphi dx = \int_D \nabla u \nabla w \varphi dx + < \Delta w, u \varphi >_{H^{-1}(D) \times H_0^1(D)},$$

formally we write

$$(4.14) \qquad \int_D \nabla u \nabla (\varphi w) dx + \int_D u \varphi w d\mu = \int_D f \varphi w dx,$$

where μ is the Borel measure defined by

$$(4.15) \qquad \mu(B) = \begin{cases} +\infty & \text{if } \text{Cap}(B \cap \{w = 0\}) > 0 \\ \int_B \frac{1}{w} d\nu & \text{if } \text{Cap}(B \cap \{w = 0\}) = 0. \end{cases}$$

Here $\nu = \Delta w + 1 \geq 0$ in $\mathcal{D}'(D)$ is a non-negative Radon measure belonging to $H^{-1}(D)$.

This formal calculus needs several rigorous proofs for which we refer the reader to [100]. We recall here the following facts

i) u vanishes where w vanishes and $u \in H_0^1(D) \cap L^2(D, \mu)$;

ii) the set $\{\varphi w : \varphi \in C_0^\infty(D)\}$ is dense in $H_0^1(D) \cap L^2(D, \mu)$;

iii) $w \in \mathcal{K} := \{w \in H_0^1(D) : w \geq 0, -\Delta w \leq 1 \text{ in } D\}$;

iv) there exists a one to one mapping between $\mathcal{K}$ and $\mathcal{M}_0(D)$ given by $w \mapsto \mu$ where μ is defined by (4.15);

v) for every $w \in \mathcal{K}$ and every $\varepsilon > 0$, there exists an open set $A \subseteq D$ such that $\|w - w_A\|_{L^2(D)} \leq \varepsilon$.

Assertions i) and ii) give full sense to equation (4.14). Assertion v) proves that the family of open sets is dense in the family of relaxed domains which are identified with measures of $\mathcal{M}_0(D)$.

Remark 4.3.10 The same construction of relaxed domains can be performed for the p-Laplacian in $W_0^{1,p}(D)$ for $1 < p < +\infty$ (see [100]).

The example of Cioranescu and Murat. In this example, we construct a sequence of open sets which is γ-convergent to an element of $\mathcal{M}_0(D)$ which is not a quasi-open set. Let D be an open set contained in the unit square of $\mathbb{R}^2$, $S =]0, 1[\times]0, 1[$.

We consider the sequence of sets

$$A_n = D \setminus \bigcup_{i,j=0}^{n} \overline{B}_{(i/n,j/n),r_n},$$

where $r_n = ce^{-n^2}$, $c > 0$ being a fixed positive constant. Let us denote by u_n the solution of (4.13) on A_n. For a subsequence, still denoted with the same indices we can suppose that $u_n \rightharpoonup u$ weakly in $H_0^1(S)$.

Instead of working with the functions w_n used for finding the general form of a relaxed problem, in this particular case is more convenient to introduce the following functions $z_n \in H^1(S)$:

(4.16)

$$z_n = \begin{cases} 0 & \text{on } \bigcup_{i,j=0}^{n} \overline{B}_{(i/n,j/n),r_n} \\[2mm] \dfrac{\log\sqrt{x^2+y^2} + cn^2}{cn^2 - \log 2n} & \text{on } \bigcup_{i,j=0}^{n} \overline{B}_{(i/n,j/n),1/2n} \setminus \bigcup_{i,j=0}^{n} \overline{B}_{(i/n,j/n),r_n} \\[2mm] 1 & \text{on } S \setminus \bigcup_{i,j=0}^{n} \overline{B}_{(i/n,j/n),\frac{1}{2n}} \end{cases}$$

We notice the following facts:

- $0 \le z_n \le 1$

- $\int_S |\nabla z_n|^2 dx \to 0$ for $n \to \infty$, hence z_n converges strongly in $H^1(S)$ to a constant function. Computing the limit of $\int_S z_n dx$ we find that this constant is equal to 1.

Let $\varphi \in C_0^\infty(D)$. Then $z_n \varphi \in H_0^1(A_n)$, thus we can take $z_n \varphi$ as a test function for equation (4.13) on A_n:

$$\int_D \nabla u_n \nabla z_n \varphi dx + \int_D \nabla u_n \nabla \varphi z_n dx = \int_D f \varphi z_n dx.$$

The second and third terms of this identity converge to $\int_D \nabla u \nabla \varphi dx$ and $\int_D f \varphi dx$, respectively. For the first term, the Green formula gives

$$\int_D \nabla u_n \nabla z_n \varphi dx = \sum_{i,j=0}^{n} \int_{\partial B_{(i/n,j/n),1/2n}} u_n \frac{\partial z_n}{\partial n} \varphi d\sigma - \int_D u_n \nabla z_n \nabla \varphi dx.$$

The boundary term on $\partial B_{(i/n,j/n),e^{-cn^2}}$ does not appear since z_n vanishes on it. The last term of this identity converges to 0 when $n \to \infty$.
  We have

$$\sum_{i,j=0}^{n} \int_{\partial B_{(i/n,j/n),1/2n}} u_n \frac{\partial z_n}{\partial n} \varphi d\sigma = \sum_{i,j=0}^{n} \int_{\partial B_{(i/n,j/n),1/2n}} \frac{2n}{cn^2 - \log(2n)} u_n \varphi d\sigma$$

$$= \frac{2n^2}{cn^2 - \log(2n)} \sum_{i,j=0}^{n} \int_{\partial B_{(i/n,j/n),1/2n}} \frac{1}{n} u_n \varphi d\sigma.$$

Let us denote by $\mu_n \in H^{-1}(S)$ the distribution defined by

$$\langle \mu_n, \psi \rangle_{H^{-1}(S) \times H_0^1(S)} = \sum_{i,j=0}^{n} \int_{\partial B_{(i/n,j/n),1/2n}} \frac{1}{n} \psi d\sigma.$$

We prove that μ_n converges strongly in $H^{-1}(S)$ to πdx. Indeed, we introduce the functions $v_n \in H_0^1(S)$ defined by $v_n = 0$ on $S \setminus \bigcup B_{(i/n,j/n),1/2n}$, $\Delta v_n = 4$ in $\bigcup B_{(i/n,j/n),1/2n}$, therefore $\partial v_n / \partial n = 1/n$ on $\bigcup \partial B_{(i/n,j/n),1/2n}$.
  We notice that $v_n \longrightarrow 0$ strongly in $H_0^1(S)$, therefore $\Delta v_n \longrightarrow 0$ strongly in $H^{-1}(S)$. But,

$$\langle -\Delta v_n, \psi \rangle_{H^{-1}(S) \times H_0^1(S)} = \sum_{i,j=0}^{n} \int_{B_{(i/n,j/n),1/2n}} \nabla v_n \nabla \psi dx$$

$$= \sum_{i,j=0}^{n} \int_{\partial B_{(i/n,j/n),1/2n}} \frac{1}{n} \psi d\sigma - \sum_{i,j=0}^{n} \int_{B_{(i/n,j/n),1/2n}} 4\psi dx.$$

Passing to the limit and using the fact that $1_{\bigcup B_{(i/n,j/n),1/2n}} \rightharpoonup \frac{\pi}{4} 1_S$ weakly in L^2 we get that

$$\mu_n \xrightarrow{H^{-1}(S)} \pi dx.$$

Consequently, the equation satisfied by $u \in H_0^1(S)$ is

$$\forall \varphi \in C_0^\infty(D) \quad \int_D \nabla u \nabla \varphi dx + \frac{2\pi}{c} \int_D u\varphi dx = \int_D f\varphi dx.$$

The following result will be extensively used in the next chapters. A detailed proof can be found in [64] for $p = 2$ and in [100] for $p \neq 2$.

Lemma 4.3.11 *Let (A_n) be a sequence of quasi-open subsets of D and let $w \in W_0^{1,p}(D)$ be a function such that $w_{A_n} \rightharpoonup w$ weakly in $W_0^{1,p}(D)$. Let $u_n \in W_0^{1,p}(D)$ be such that $u_n = 0$ p-q.e. on $D \setminus A_n$ and suppose that $u_n \rightharpoonup u$ in $W_0^{1,p}(D)$. Then $u = 0$ p-q.e. on $\{w = 0\}$.*

Proof This follows from the fact that the functionals

$$F_n(u) = \int_B |\nabla u|^p dx + \chi_{W_0^{1,p}(\Omega_n)}(u)$$

Γ-converge in $L^p(B)$-*strong* to the functional

$$F(u) = \int_B |\nabla u|^p dx + \int_B |u|^p d\mu$$

μ being the Radon measure defined by

$$(4.17) \qquad \mu(A) = \begin{cases} +\infty & if \; \; \mathrm{cap}_p(A \cap \{w = 0\}) > 0 \\ \int_A \frac{d\nu}{w^{p-1}} & if \; \; \mathrm{cap}_p(A \cap \{w = 0\}) = 0. \end{cases}$$

Here $\chi_{W_0^{1,p}(\Omega)}(u) = 0$ if $u \in W_0^{1,p}(\Omega)$ and $+\infty$ if not. Notice that (4.17) is the non-linear version of (4.15). ∎

4.4 Necessary conditions of optimality

In this section we consider the shape optimization problem

$$(4.18) \qquad \min\left\{ \int_D j(x, u_A)\, dx \; : \; A \text{ open subset of } D \right\}$$

where we denoted by u_A the unique solution of the Dirichlet problem

$$-\Delta u = f \text{ in } A, \qquad u \in H_0^1(A).$$

Here D is a bounded open subset of $\mathbb{R}^N$ ($N \geq 2$), $f \in L^2(D)$, and the integrand $j(x, s)$ is supposed to be a Carathéodory function such that

$$(4.19) \qquad |j(x, s)| \leq a(x) + c|s|^2$$

for suitable $a \in L^1(D)$ and $c \in \mathbb{R}$.

As seen in the Sections 3.5, 3.6, 4.3 the relaxed form of the shape optimization problem above involves measures of $\mathcal{M}_0(D)$ as relaxed controls, and takes the form

$$(4.20) \qquad \min\left\{ \int_D j(x, u_\mu)\, dx \; : \; \mu \in \mathcal{M}_0(D) \right\}$$

where we denoted by u_μ the unique solution of the relaxed Dirichlet problem

$$u \in X_\mu(D), \qquad -\Delta u + \mu u = f \text{ in } X_\mu'(D).$$

Remark 4.4.1 *By using the Sobolev embedding theorem, it is easy to see that it is possible to replace the growth condition (4.19) by the weaker one*

$$(4.21) \qquad |j(x, s)| \leq a(x) + c|s|^p$$

with $a \in L^1(D)$, $c \in \mathbb{R}$, and $p < 2N/(N-2)$.

We have already seen examples which show that the original problem (4.18) may have no solution; on the other hand, the relaxed optimization problem (4.20) always admits a solution, as shown in the abstract scheme of Section 3.5. Our goal is now to obtain some necessary conditions of optimality for the solutions μ of the relaxed optimization problem (4.20). They will be obtained by evaluating the cost functional on a family μ_ε of perturbations of μ and by computing the limit

$$\lim_{\varepsilon \to 0^+} \frac{J(u_{\mu_\varepsilon}) - J(u_\mu)}{\varepsilon} \, .$$

In what follows we assume for simplicity that the function $j(x, \cdot)$ is continuously differentiable and that its differential verifies the growth condition

$$|j_s(x, s)| \leq a_1(x) + c_1|s|$$

for suitable $a_1 \in L^2(D)$ and $c_1 \in \mathbb{R}$.

The first perturbation we consider is of the form $\mu_\varepsilon = \mu + \varepsilon \phi \mathcal{H}^N$ where ϕ is a nonnegative function belonging to $L^\infty(D)$. If (u, μ) is an optimal pair of the relaxed optimization problem and $u_\varepsilon = R_{\mu_\varepsilon}(f)$, proceeding as in [63] we obtain

$$\frac{dJ(u_\varepsilon)}{d\varepsilon}\Big|_{\varepsilon=0} = -\int_D j_s(x, u) R_\mu(\phi u) \, dx.$$

On the other hand, the optimality of μ gives that the derivative above has to be nonnegative, so that we obtain

$$\int_D j_s(x, u) R_\mu(\phi u) \, dx \leq 0 \qquad \forall \phi \in L^\infty, \ \phi \geq 0.$$

By the symmetry of the resolvent operator R_μ we can also write

$$\int_D R_\mu\big(j_s(x, u)\big) \phi u \, dx \leq 0 \qquad \forall \phi \in L^\infty, \ \phi \geq 0$$

which gives, since ϕ is arbitrary,

$$R_\mu\big(j_s(x, u)\big) u \leq 0 \qquad \text{a.e. in } D.$$

It is now convenient to introduce the adjoint state equation

(4.22) $v \in X_\mu(D), \qquad -\Delta v + \mu v = j_s(x, u) \text{ in } X'_\mu(D)$

so that the optimality condition above reads

$$uv \leq 0 \qquad \text{a.e. in } D.$$

Noticing that u and v are finely continuous q.e. in D their product uv is still finely continuous q.e. in D, and since nonempty finely open sets have positive Lebesgue measure we obtain the following necessary condition of optimality.

Proposition 4.4.2 *If (u, μ) is an optimal pair of the relaxed optimization problem (4.20) and if v denotes the solution of the adjoint state equation (4.22), then we have*

$$(4.23) \qquad\qquad uv \leq 0 \qquad q.e. \ in \ D.$$

We consider now another kind of perturbation of an optimal measure μ by taking the family of measures $\mu_\varepsilon = (1 - \varepsilon)\mu$ that, for $\varepsilon < 1$, still belong to the class $\mathcal{M}_0(D)$. Again, denoting by $u_\varepsilon = R_{\mu_\varepsilon}(f)$ the optimal state related to μ_ε, proceeding as in [63] we obtain

$$\frac{dJ(u_\varepsilon)}{d\varepsilon}\Big|_{\varepsilon=0} = \int_D j_s(x, u) R_\mu(\mu u) \, dx.$$

The optimality of μ gives that the derivative above has to be nonnegative, so that we obtain

$$\int_D j_s(x, u) R_\mu(\mu u) \, dx \geq 0$$

and, again by the symmetry of the resolvent operator R_μ we have

$$\int_D R_\mu\big(j_s(x, u)\big) u \, d\mu \geq 0.$$

On the other hand, the optimality condition obtained in Proposition 4.4.2 gives that the product $u R_\mu\big(j_s(x, u)\big)$ is less than or equal to zero q.e. in D, hence μ-a.e. in D, which implies the following second necessary condition of optimality.

Proposition 4.4.3 *If (u, μ) is an optimal pair of the relaxed optimization problem (4.20) and if v denotes the solution of the adjoint state equation (4.22), then we have*

$$(4.24) \qquad\qquad uv = 0 \qquad \mu\text{-}a.e. \ in \ D.$$

In order to obtain further necessary conditions of optimality it is convenient to introduce, for every finely open subset A of $\mathbb{R}^N$, a boundary measure ν_A, carried by the fine boundary $\partial^* A$. If we denote by w_A the unique solution of the Dirichlet problem

$$-\Delta w_A = 1 \text{ in } A, \qquad w_A \in H_0^1(A),$$

then the following theorem gives the existence of ν_A.

Theorem 4.4.4 *There exists a unique nonnegative measure ν_A belonging to $H^{-1}(\mathbb{R}^N)$ and such that*

$$(4.25) \qquad\qquad -\Delta w_A + \nu_A = 1_{cl^*A} \ in \ H^{-1}(\mathbb{R}^N),$$

where cl^*A denotes the fine closure of A. Moreover, we have that ν_A is carried by ∂^*A $\left(i.e. \ \nu_A(\mathbb{R}^N \setminus \partial^*A) = 0\right)$, and

$$\nu_A(\partial^*A) = \mathcal{H}^N(cl^*A).$$

For the proof of theorem above we refer to [63].

Example 4.4.5 *If A is a smooth domain then $\partial^*A = \partial A$ and the solution w_A is smooth up to the boundary. Using (4.25) an integration by parts gives*

$$\int_{\mathbb{R}^N} v \, d\nu_A = -\int_{\partial A} v \frac{\partial w_A}{\partial n} d\mathcal{H}^{N-1} \qquad \forall v \in H^1(\mathbb{R}^N)$$

where n is the outer unit normal vector to A. Thus

$$\nu_A = -\frac{\partial w_A}{\partial n} \mathcal{H}^{N-1} \llcorner \partial A.$$

The measure $\bar{\nu}_A$ above allows us to give a weak definition of the normal derivative for the solution u of a relaxed state equation

$$(4.26) \qquad\qquad -\Delta u + \mu u = f \ in \ X'_\mu(D), \qquad u \in X_\mu(D),$$

where $f \in L^2(D)$ and $\mu \in \mathcal{M}_0(D)$. We denote by $A = A_\mu$ the set of finiteness of μ, as defined in the previous section, and by ν_A the boundary measure defined above. The following result holds (see [63] for the proof).

Proposition 4.4.6 *There exists a unique $\alpha \in L^2(D, \nu_A)$ such that*

$$(4.27) \qquad\qquad -\Delta u + \mu u + \alpha \nu_A = f 1_{cl^*A} \ in \ H^{-1}(D).$$

Moreover we have

$$\int_D \alpha^2 \, d\nu_A \leq \int_D f^2 \, dx$$

and $\alpha \geq 0$ ν_A-a.e. in D whenever $f \geq 0$ a.e. in D.

Example 4.4.7 Let A be a smooth domain and let $\mu = a(x)\mathcal{H}^N \llcorner A$, with $a \in L^\infty(A)$. Then equation (4.26) simply reads

$$-\Delta u + a(x)u = f \text{ in } H^{-1}(A) \qquad u \in H^1_0(A),$$

and $u \in H^2(A)$, so that $\partial u/\partial n \in L^2(\partial A, \mathcal{H}^{N-1})$. Using (4.27) and integrating by parts we obtain

$$\int_D v\alpha \, d\nu_A = -\int_{D\cap\partial A} v\frac{\partial u}{\partial n} \, d\mathcal{H}^{N-1} \qquad \forall v \in H^1(\mathbb{R}^N).$$

Therefore

$$\alpha\nu_A = -\frac{\partial u}{\partial n}\mathcal{H}^{N-1}\llcorner D \cap \partial A.$$

Since by Example 4.4.5 we have $\nu_A = -\frac{\partial w_A}{\partial n}\mathcal{H}^{N-1}\llcorner\partial A$, we finally deduce

$$\alpha = \frac{\partial u}{\partial n}\bigg/\frac{\partial w_A}{\partial n} \qquad \mathcal{H}^{N-1} \text{ a.e. on } D \cap \partial A.$$

Note that by the Hopf maximum principle we have $\partial w_A/\partial n < 0$ on ∂A.

The last two necessary conditions of optimality for a solution μ of the relaxed problem (4.20) will be obtained by considering the perturbation

$$\mu_\varepsilon = \mu\llcorner A + \frac{1}{\varepsilon}\left(\frac{1}{\phi}\mathcal{H}^N\llcorner \text{int}^*S + \frac{1}{\psi}\nu_A\right)$$

where $A = A_\mu$, $S = S_\mu$, and ϕ, ψ are two positive and continuous functions up to $\overline{D}$. We give a sketch of the proof by referring to [63] for all details. We denote by u_ε the corresponding solution of

$$u_\varepsilon \in X_{\mu_\varepsilon}(D), \qquad -\Delta u_\varepsilon + \mu_\varepsilon u_\varepsilon = f \text{ in } X'_{\mu_\varepsilon}(D).$$

It is possible to show that $u_\varepsilon \to u$ strongly in $H^1_0(D)$. Moreover, by Proposition 4.4.6 there exists a unique $\alpha \in L^2(D, \nu_A)$ such that

$$-\Delta u + \mu u + \alpha\nu_A = f1_{cl^*A} \text{ in } H^{-1}(D);$$

analogously, if v is the solution of the adjoint equation

$$v \in X_\mu(D), \qquad -\Delta v + \mu v = j_s(x, u) \text{ in } X'_\mu(D)$$

there exists a unique $\beta \in L^2(D, \nu_A)$ such that

$$-\Delta v + \mu v + \beta \nu_A = j_s(x, u)1_{cl^*A} \text{ in } H^{-1}(D).$$

Then, for every $g \in L^2(D)$ it is possible to compute the limit

$$\lim_{\varepsilon \to 0} \int_D \frac{u_\varepsilon - u}{\varepsilon} g \, dx$$

in terms of the function α, β introduced above, and, by using the fact that ϕ, ψ are arbitrary, we obtain the further necessary conditions of optimality:

(4.28)
$$\begin{aligned} f(x)j_s(x, 0) &\geq 0 \qquad \text{for a.e. } x \in int^*S \\ \alpha\beta &\geq 0 \qquad \nu_A\text{-a.e. on } D. \end{aligned}$$

Summarizing, the four optimality conditions we have obtained are:

(4.29)
$$\begin{aligned} uv &\leq 0 \qquad \text{q.e. in } D; \\ uv &= 0 \qquad \mu\text{-a.e. in } D; \\ f(x)j_s(x, 0) &\geq 0 \qquad \text{for a.e. } x \in int^*S; \\ \alpha\beta &\geq 0 \qquad \nu_A\text{-a.e. on } D. \end{aligned}$$

Example 4.4.8 It is interesting to rewrite the conditions above in the case when the optimal measure μ has the form

$$\mu = a(x)\mathcal{H}^N \llcorner A + \infty_{D \setminus A}$$

with $a \in L^\infty(D)$, $a(x) \geq 0$ for a.e. $x \in D$, and A is an open subset of D with a smooth boundary. In this case the optimality conditions above become:

(4.30)
$$\begin{aligned} uv &\leq 0 \qquad \text{q.e. on } A; \\ uv &= 0 \qquad \text{a.e. on } \{x \in A : a(x) > 0\}; \\ f(x)j_s(x, 0) &\geq 0 \qquad \text{for a.e. } x \in D \setminus A; \\ (\partial u/\partial n)(\partial v/\partial n) &\geq 0 \qquad \mathcal{H}^{N-1}\text{-a.e. on } D \cap \partial A. \end{aligned}$$

Since the boundary of A has been assumed smooth, the optimal state u and its adjoint state v both belong to the Sobolev space $H^2(A)$; hence the last condition can be written in the stronger form

$$(\partial u/\partial n)(\partial v/\partial n) = 0 \qquad \mathcal{H}^{N-1}\text{-a.e. on } D \cap \partial A.$$

Indeed, using the fact that $uv \leq 0$ on A, this follows by considering the one-dimensional functions $t \mapsto u\big(x + tn(x)\big)$ and $t \mapsto v\big(x + tn(x)\big)$ which are continuously differentiable in a neighborhood of $t = 0$ for $\mathcal{H}^{N-1}$-a.e. point $x \in D \cap \partial A$.

Specializing the conditions above to the particular case when $a \equiv 0$, which means that the original shape optimization problem has a classical solution, we obtain:

$$
(4.31) \qquad
\begin{array}{ll}
uv \leq 0 & \text{q.e. on } A; \\
f(x)j_s(x,0) \geq 0 & \text{for a.e. } x \in D \setminus A; \\
(\partial u/\partial n)(\partial v/\partial n) = 0 & \mathcal{H}^{N-1}\text{-a.e. on } D \cap \partial A.
\end{array}
$$

Remark 4.4.9 In general, for a shape optimization problem of the form

$$
\min_{\Omega \in \mathcal{U}_{ad}} J(\Omega),
$$

which has a classical solution $\Omega^* \in \mathcal{U}_{ad}$, one can write two types of necessary optimality conditions that we briefly describe below.

Using the shape derivative. For an admissible vector field V one computes the shape derivative

$$
dJ(\Omega^*; V) = \lim_{t \to 0} \frac{J\big((Id + tV)\Omega^*\big) - J(\Omega^*)}{t}.
$$

Of course, the vector field V has to be chosen in such a way that $(Id + tV)\Omega^* \in \mathcal{U}_{ad}$, or to use a Lagrange multiplier. The optimality condition writes

$$
dJ(\Omega^*; V) \geq 0.
$$

Usually, the computation of the shape derivative requires that Ω^* is smooth enough (C^2 for example); the regularity of the optimum Ω^* is, in general, very difficult to prove. Particular attention has to be given to the case when $\mathcal{U}_{ad}$ consists of convex sets, since the convexity constraint is "unstable" to small variations of the boundary. We refer the reader to [107], [135], [176] for detailed discussions of the shape derivative.

Using the topological derivative. For every $x_0 \in \Omega^*$, one computes the following asymptotic development

$$
J(\Omega^* \setminus \overline{B}_{x_0,\varepsilon}) = J(\Omega^*) + g(x_0)f(\varepsilon) + o(f(\varepsilon)),
$$

where $f(\varepsilon) > 0$ is such that $\lim_{\varepsilon \to 0} f(\varepsilon) = 0$. The optimality condition writes $g(x_0) \geq 0$. We refer to [123], [175] for a detailed discussion of the topological derivative and for several applications to concrete problems.

4.5   Boundary variation

In this section we study the continuity of the mapping

$$\Omega \longrightarrow u_\Omega$$

where u_Ω is the weak variational solution of the following Dirichlet problem on Ω

(4.32)
$$\left\{ \begin{array}{l} -\Delta u_\Omega = f \\ u_\Omega \in H_0^1(\Omega) \end{array} \right.$$

Here, $\Omega \subseteq D$ is an open subset of a bounded design region D of $\mathbb{R}^N$, with $N \geq 2$ and $f \in H^{-1}(D)$ is a fixed distribution. Since $u_\Omega \in H_0^1(\Omega)$, by extension with zero we can suppose that $u_\Omega \in H_0^1(D)$.

When speaking about "shape continuity" one has to endow the space of open sets with a topology. Obviously, continuity holds if the family of open sets is endowed with a rather strong topology. Such a strong topology is for example given by the following distance

$$d(\Omega_1, \Omega_2) = \inf\{\|T - Id\| + \|T^{-1} - Id\|\},$$

the infimum being taken over all C^1-diffeomorphisms T of the space $\mathbb{R}^N$ with $T(\Omega_1) = \Omega_2$ (see [176]).

If one considers a weaker topology on the space of admissible open sets, the continuity may not hold. Nevertheless, the interests to consider a weak topology is high when dealing with shape optimization problems. Indeed, in view of applying the direct methods of the calculus of variations, one needs compactness, and this is easier to be obtained working with weak topologies.

Therefore one has to look for a kind of equilibrium: a topology weak enough in order to have compactness but strong enough in order to get continuity. This purpose is not easy to be attained; what we can do, is to search weak geometrical or topological constraints which would make a class of domains compact (in the chosen topology) and which are sufficient to give shape continuity.

Notice that equation (4.32) is set in a very simple case: homogeneous Dirichlet boundary conditions for the Laplace operator. We chose this easy setting only to simplify the proofs and to avoid heavy notations. At the end of the section we discuss how all results can be extended to non-linear elliptic operators, non-homogeneous boundary conditions and evolution equations.

Roughly speaking, the shape continuity of the solution does not depend "so much" on the operator and on the right hand side f, on the other hand it "strongly" depends on the sequence of domains.

Two points of view may occur when dealing with the shape continuity of the solution of a PDE. Either we suppose that an open set Ω is given and (Ω_n) is a perturbation of it or, when we search compactness, that (Ω_n) is given and we search an open set Ω such that (for a subsequence denoted using the same indices) we have that $u_{\Omega_n} \to u_\Omega$. In the latter case, one has to endow the space of open sets with a suitable topology τ. The construction of Ω might be done using compactness properties of this topology. The shape continuity of the solution is deduced from the geometrical constraints imposed on Ω_n. The role of these constraints is double: on one side they make the topological space $(\mathcal{A}, \tau)$ compact, and on the other side they provide the γ-convergence of any τ-convergent sequence (here $\mathcal{A}$ denotes the class of admissible domains).

In this section we mainly deal with the second point of view. The first one will be rapidly discussed at the end of the next section; it will be deduced as a particular case.

How to prove shape continuity. Suppose that $(\Omega_n)_n$ is a sequence of open sets contained in a bounded design region D. It is said that the sequence of spaces $(H_0^1(\Omega_n))$ converges in the sense of Mosco to the space $H_0^1(\Omega)$ if the following conditions are satisfied:

1. For all $\phi \in H_0^1(\Omega)$ there exists a sequence $\phi_n \in H_0^1(\Omega_n)$ such that ϕ_n converges strongly in $H_0^1(D)$ to ϕ.

2. For any sequence $\phi_{n_k} \in H_0^1(\Omega_{n_k})$ weakly convergent in $H_0^1(D)$ to a function ϕ we have $\phi \in H_0^1(\Omega)$.

For every open set $\Omega \subseteq D$ we denote by $P_{H_0^1(\Omega)}$ the orthogonal projection of $H_0^1(D)$ onto $H_0^1(\Omega)$. By R_Ω we denote the resolvent operator $R_\Omega : L^2(D) \to L^2(D)$ defined by $R_\Omega(f) = u_{\Omega,f}$.

Lemma 4.5.1 *Let Ω be an open subset of D. There exists a constant M depending only on f and D such*

$$\|u_{\Omega,f}\|_{H_0^1(D)} \leq M.$$

Proof Take $u_{\Omega,f}$ as test function in the weak formulation of the equation and apply the Cauchy inequality. ∎

Proposition 4.5.2 *Let $(\Omega_n)_n$ and Ω be open subsets of D. The following assertions are equivalent.*

1. *for every $f \in H^{-1}(D)$ we have $u_{\Omega_n,f} \longrightarrow u_{\Omega,f}$ strongly in $H_0^1(D)$;*

2. *for $f \equiv 1$ we have $u_{\Omega_n,1} \longrightarrow u_{\Omega,1}$ strongly in $H_0^1(D)$;*

3. *$H_0^1(\Omega_n)$ converges in the sense of Mosco to $H_0^1(\Omega)$;*

4. *$G(\Omega_n, \cdot) \xrightarrow{\Gamma} G(\Omega, \cdot)$ in $L^2(D)$, where $G(A, \cdot)$ are the associated energy functionals defined in (3.26) for $f \equiv 0$;*

5. *For every $u \in H_0^1(D)$ the sequence $(P_{H_0^1(\Omega_n)}(u))$ converges strongly in $H_0^1(D)$ to $P_{H_0^1(\Omega)}(u)$.*

6. *R_{Ω_n} converges in the operator norm of $\mathcal{L}(L^2(D))$ to R_Ω.*

Proof 1) $\Longrightarrow$ 2) is obvious.

2) $\Longrightarrow$ 1) Let $f \in L^\infty(D), f \geq 0$. By lemma 4.5.1, for a subsequence we have

$$u_{\Omega_{n_k},f} \overset{H_0^1(D)}{\rightharpoonup} u.$$

From the maximum principle

$$0 \leq u_{\Omega_{n_k},f} \leq \|f\|_\infty u_{\Omega_{n_k},1},$$

hence passing to the limit

$$0 \leq u \leq \|f\|_\infty u_{\Omega,1},$$

therefore $u \in H_0^1(\Omega)$. Let now $\varphi \in \mathcal{D}(\Omega)$; there exists $\alpha > 0$ such that

$$0 \leq |\varphi| \leq \alpha u_{\Omega,1}.$$

We define the sequence $\varphi_n = \varphi_n^+ - \varphi_n^-$, where

$$\varphi_n^+ = \min\{\varphi^+, \alpha u_{\Omega,1}\} \quad \varphi_n^- = \min\{\varphi^-, \alpha u_{\Omega,1}\}.$$

On one side we have that $\varphi_n \in H_0^1(\Omega_n)$, and on the other side $\varphi_n \longrightarrow \varphi$ strongly in $H_0^1(D)$. Writing

$$\int_D \nabla u_{\Omega_{n_k},f} \nabla \varphi_{n_k} dx = \int_D f\varphi_n dx$$

and passing to the limit as $n \to \infty$ we get

$$\int_D \nabla u \nabla \varphi \, dx = \int_D f \varphi \, dx.$$

Consequently, $u = u_{\Omega,f}$. The convergence is strong in $H_0^1(D)$ since the norms also converge. Moreover, we get the convergence of the whole sequence from the uniqueness of the limit $u = u_{\Omega,f}$.

By the linearity of the equation and the density of $L^\infty(D)$ in $H^{-1}(D)$, the proof of point 1) is achieved.

1) $\Longleftrightarrow$ 5) Let $u \in H_0^1(D)$, and denote $f := -\Delta u \in H^{-1}(D)$. Then $P_{H_0^1(\Omega_n)}(u) = u_{\Omega_n,f}$, the equivalence between 1) and 5) being obvious.

5) $\Longrightarrow$ 4) Let $u_n \longrightarrow u$ strongly in $L^2(D)$. In order to prove that $G(\Omega, u) \leq \liminf_{n \to \infty} G(\Omega_n, u_n)$ we can suppose that $\liminf_{n \to \infty} G(\Omega_n, u_n) < +\infty$. The inequality would then follow as soon as $u \in H_0^1(\Omega)$. For that it is enough to prove that $u = P_{H_0^1(\Omega)}(u)$. For every $\varphi \in H_0^1(D)$ we have

$$
\begin{aligned}
(u, \varphi)_{H_0^1(D) \times H_0^1(D)} &= \lim_{n \to \infty} (u_n, \varphi)_{H_0^1(D) \times H_0^1(D)} \\
&= \lim_{n \to \infty} (P_{H_0^1(\Omega_n)} u_n, \varphi)_{H_0^1(D) \times H_0^1(D)} \\
&= \lim_{n \to \infty} (u_n, P_{H_0^1(\Omega_n)} \varphi)_{H_0^1(D) \times H_0^1(D)}.
\end{aligned}
$$

Using 5) and the pairing (weak,strong)-convergence we get

$$(u, \varphi)_{H_0^1(D) \times H_0^1(D)} = (u, P_{H_0^1(\Omega)} \varphi)_{H_0^1(D) \times H_0^1(D)} = (P_{H_0^1(\Omega)} u, \varphi)_{H_0^1(D) \times H_0^1(D)},$$

hence $u = P_{H_0^1(\Omega)} u$.

Let now $u \in H_0^1(\Omega)$. We define $u_n := P_{H_0^1(\Omega_n)} u \in H_0^1(\Omega_n)$ and get $u_n \longrightarrow u$ strongly in $H_0^1(D)$. Then $G(\Omega, u) = \lim_{n \to \infty} G(\Omega_n, u_n)$.

4) $\Longrightarrow$ 3) Let $u \in H_0^1(\Omega)$. From the Γ-convergence, there exits $(u_n)_n$ such that $u_n \in H_0^1(\Omega_n)$ with $u_n \longrightarrow u$ strongly in $L^2(D)$ and $G(\Omega, u) = \lim_{n \to \infty} G(\Omega_n, u_n)$. This means that $u_n \in H_0^1(\Omega_n)$ and $u_n \longrightarrow u$ strongly in $H_0^1(D)$, hence the first Mosco condition is satisfied.

For the second Mosco condition, let $u_{n_k} \in H_0^1(\Omega_{n_k})$ such that $u_{n_k} \rightharpoonup u$ weakly in $H_0^1(D)$. Hypothesis 4) gives $G(\Omega, u) \leq \liminf_{n \to \infty} G(\Omega_{n_k}, u_{n_k}) < +\infty$, i.e. $u \in H_0^1(\Omega)$.

3) $\Longrightarrow$ 1) Let $\varphi \in H_0^1(\Omega)$. From the first Mosco condition there exists $\varphi_n \in H_0^1(\Omega_n)$ such that $\varphi_n \longrightarrow \varphi$ strongly in $H_0^1(D)$. On the other hand, for

a subsequence we have

$$u_{\Omega_{n_k},f} \overset{H_0^1(D)}{\rightharpoonup} u,$$

and from the second Mosco condition $u \in H_0^1(\Omega)$. Writing the following chain of equalities we get $u = u_{\Omega,f}$:

$$\int_D \nabla u \nabla \varphi dx = \lim_{n\to\infty} \int_D \nabla u_{\Omega_n,f} \nabla \varphi_n dx$$

$$= \lim_{n\to\infty} \langle f, \varphi_n \rangle_{H^{-1}(D) \times H_0^1(D)} = \langle f, \varphi \rangle_{H^{-1}(D) \times H_0^1(D)}.$$

Classical arguments now give

$$u_{\Omega_n,f} \overset{H_0^1(D)}{\longrightarrow} u_{\Omega,f}.$$

3) $\implies$ 6) We have

$$|R_{\Omega_n} - R_\Omega|_{\mathcal{L}(L^2(\mathbb{R}^N))} = \sup_{\|f\|_{L^2(D)} \leq 1} \|R_{\Omega_n}(f) - R_\Omega(f)\|_{L^2(D)}.$$

Supposing that $f_n \in L^2(D)$ is such that $\|f_n\|_{L^2(D)} \leq 1$ and

$$|R_{\Omega_n} - R_\Omega|_{\mathcal{L}(L^2(\mathbb{R}^N))} \leq \|R_{\Omega_n}(f_n) - R_\Omega(f_n)\|_{L^2(D)} + \frac{1}{n},$$

we can assume for a subsequence (still denoted using the same indices) that $f_n \rightharpoonup f$ weakly in $L^2(D)$. For proving 6), is enough to show that $R_{\Omega_n}(f_n)$ converges strongly in $L^2(D)$ to $R_\Omega(f)$. This is a consequence of 3) and of the compact embedding $H_0^1(D) \hookrightarrow L^2(D)$.

6) $\implies$ 2) Let $f \equiv 1$. Then $R_{\Omega_n}(f) = u_{\Omega_n,1}$ and 2) follows. ∎

Remark 4.5.3 The γ-convergence is local, i.e. $\Omega_n \overset{\gamma}{\longrightarrow} \Omega$ if and only if there exists $\delta > 0$ such that for every $x \in D$ and for every $r \in (0, \delta)$ we have that $\Omega_n \cap B_{x,r} \overset{\gamma}{\longrightarrow} \Omega \cap B_{x,r}$. This can be easily proved using the Mosco convergence and a partition of unity. Moreover the following holds.

Proposition 4.5.4 *Let us consider two sequences of open sets $X_n \overset{\gamma}{\to} X$ and $Y_n \overset{\gamma}{\to} Y$. Then $X_n \cap Y_n \overset{\gamma}{\to} X \cap Y$.*

4.6 Continuity under geometric constraints

Let D be a bounded open set. In this section we set

$$\mathcal{A} = \{\Omega : \Omega \subseteq D, \ \Omega \text{ open}\}$$

and τ the Hausdorff complementary topology on $\mathcal{A}$, given by the metric

$$d_{H^c}(\Omega_1, \Omega_2) = d(\Omega_1^c, \Omega_2^c).$$

Here d is the usual Hausdorff distance introduced in Definition 2.4.1.

Proposition 4.6.1 Properties of the Hausdorff convergence.

1. $(\mathcal{A}, H^c)$ *is a compact metric space.*

2. *If $\Omega_n \xrightarrow{H^c} \Omega$, then for all compact sets $K \subseteq \Omega$, there exists $N_K \in \mathbb{N}$ such that for every $n \geq N_K$ we have $K \subseteq \Omega_n$.*

3. *The Lebesgue measure is l.s.c. in the H^c-topology.*

4. *The number of connected components of the complement of an open set is l.s.c. in the H^c-topology.*

Proof The proof of this proposition is quite simple; we refer to [135] for further details. We notice that property 1) is a consequence of the Ascoli-Arzela theorem. ∎

Notice that the first Mosco condition is fulfilled for every sequence $\Omega_n \xrightarrow{H^c} \Omega$. Moreover, the space $(\mathcal{A}, \tau)$ is compact, even if it does not turn out to be γ-compact.

Proposition 4.6.2 *Suppose that $\Omega_n \xrightarrow{H^c} \Omega$. There exists a subsequence of $(\Omega_n)_{n \in \mathbb{N}}$, still denoted using the same indices, such that:*

$$u_{\Omega_n} \overset{H_0^1(D)}{\rightharpoonup} u$$

and

(4.33)
$$\int_\Omega \langle \nabla u, \nabla \phi \rangle dx = \langle f, \phi \rangle_{H^{-1}(D) \times H_0^1(D)}$$

for every $\phi \in H_0^1(\Omega)$.

Proof By Lemma 4.5.1 the sequence (u_{Ω_n}) is bounded in $H_0^1(D)$ so that we may assume it converges weakly to some function $u \in H_0^1(D)$. It remains to prove equality (4.33). By a density argument, we may take $\phi \in C_c^\infty(\Omega)$. Since the support of ϕ is compact and Ω_n converges in H^c to Ω, equality (4.33) is valid for u_{Ω_n}, when n is large enough. The proof is then achieved by passing to the limit as $n \to \infty$. ∎

In order to get $u = u_\Omega$ one needs only to prove that $u \in H_0^1(\Omega)$, i.e. the second Mosco condition. In order to have the second Mosco condition the geometrical constraints play a crucial role.

The following counterexample shows that, in general, H^c-convergent sequences are not γ-convergent.

Example 4.6.3 Let $\{x_1, x_2, ...\}$ be the sequence of points of rational coordinates of the square $D =]0, 1[\times]0, 1[$ in $\mathbb{R}^2$. Defining $\Omega_n = D \setminus \{x_1, x_2, ..., x_n\}$ we get that $\Omega_n \xrightarrow{H^c} \emptyset$ and $\Omega_n \xrightarrow{\gamma} D$ since $\mathrm{cap}(D \setminus \Omega_n) = 0$.

A non-exhaustive list of classes of domains in which the γ-convergence is equivalent to the H^c-convergence is the following (from the strongest constraints to the weakest ones).

- The class $\mathcal{A}_{convex} \subseteq \mathcal{A}$ of convex sets contained in D.

- The class $\mathcal{A}_{unif\ cone} \subseteq \mathcal{A}$ of domains satisfying a uniform exterior cone property (see Chenais [80], [81]), i.e. such that for every point x_0 on the boundary of every $\Omega \in \mathcal{A}_{unif\ cone}$ there is a closed cone, with uniform height and opening, and with vertex in x_0, lying in the complement of Ω.

- The class $\mathcal{A}_{unif\ flat\ cone}$ of domains satisfying a uniform flat cone condition (see Bucur, Zolésio [50]), i.e. as above, but with the weaker requirement that the cone may be flat, that is of dimension $N - 1$.

- The class $\mathcal{A}_{cap\ density} \subseteq \mathcal{A}$ of domains satisfying a uniform capacitary density condition (see [50]), i.e. such that there exist $c, r > 0$ such that for every $\Omega \in \mathcal{A}_{cap\ density}$, and for every $x \in \partial\Omega$, we have

$$\forall t \in (0, r) \quad \frac{\mathrm{cap}(\Omega^c \cap B_{x,t}, B_{x,2t})}{\mathrm{cap}(B_{x,t}, B_{x,2t})} \geq c,$$

where $B_{x,s}$ denotes the ball of radius s centered at x.

- The class $\mathcal{A}_{unif\,Wiener} \subseteq \mathcal{A}$ of domains satisfying a uniform Wiener condition (see [49]), i.e. domains satisfying for every $\Omega \in \mathcal{A}_{unif\,Wiener}$ and for every point $x \in \partial\Omega$

$$\int_r^R \frac{\operatorname{cap}(\Omega^c \cap B_{x,t}, B_{x,2t})}{\operatorname{cap}(B_{x,t}, B_{x,2t})} \frac{dt}{t} \geq g(r, R, x) \text{ for every } 0 < r < R < 1$$

where $g : (0,1) \times (0,1) \times B \to \mathbb{R}_+$ is fixed, such that for every $R \in (0,1)$ $\lim_{r \to 0} g(r, R, x) = +\infty$ locally uniformly on x.

Another interesting class, which is only of topological type and is not contained in any of the previous ones, was given by Šverák [177] and consists in the following.

- For $N = 2$, the class of all open subsets Ω of D for which the number of connected components of $\overline{D} \setminus \Omega$ is uniformly bounded.

In fact, we shall see that this last constraint is strongly related to a capacity density type constraint: in two dimensions, any curve has a strictly positive capacity.

Roughly speaking, the following inclusions can be established:

$$\mathcal{A}_{convex} \subseteq \mathcal{A}_{unif\,cone} \subseteq \mathcal{A}_{unif\,flat\,cone} \subseteq \mathcal{A}_{cap\,density} \subseteq \mathcal{A}_{unif\,Wiener},$$

hence it would be enough to prove that the γ-convergence is equivalent to the H^c-convergence only in $\mathcal{A}_{unif\,Wiener}$. The shape continuity under a uniform Wiener criterion was first observed by Frehse [120]. The proof of the continuity under the uniform Wiener criterion is slightly more technical than the continuity under capacity density condition. This is the main reason for which we prove in the sequel the continuity result only in $\mathcal{A}_{cap\,density}$ which is based on a uniform Holder estimation of the solutions (for the right hand side $f \equiv 1$) on the moving domain. A uniform Wiener condition is in some sense the weakest reasonable constraint to obtain a continuity result in the Hausdorff complementary topology; it is based on a local equi-continuity-like property of the solutions on the moving domain.

The last part of this chapter is devoted to find necessary and sufficient conditions for the shape continuity. In order to introduce the reader to non-linear equations, the results of the last section are presented for the p-Laplacian (with $1 < p < +\infty$). A careful reading of this section gives an idea to the reader on how to prove shape continuity under uniform Wiener criterion.

Definition 4.6.4 *For $r, c > 0$ it is said that an open set Ω has the (r,c) capacity density condition if*

$$(4.34) \qquad \forall x \in \partial\Omega, \forall t \in (0, r) \quad \frac{\mathrm{cap}(\Omega^c \cap B_{x,t}, B_{x,2t})}{\mathrm{cap}(B_{x,t}, B_{x,2t})} \geq c.$$

The class of open subsets of D having the (r,c) capacity density condition is denoted $\mathcal{O}_{c,r}(D)$.

We recall the following lemmas from [136] in the nonlinear setting ($1 < p < +\infty$, $q = \frac{p}{p-1}$).

Lemma 4.6.5 *Suppose that Ω is bounded. Let $\theta \in W^{1,p}(\Omega) \cap C(\overline{\Omega})$ and let h be the unique p-harmonic function in Ω with $\theta - h \in W_0^{1,p}(\Omega)$. If $x_0 \in \partial\Omega$, then for every $0 < r \leq R$ we have*

$$osc(h, \Omega \cap B_{x_0,r}) \leq osc(\theta, \partial\Omega \cap \overline{B}_{x_0,2R}) + osc(\theta, \partial\Omega)exp(-cw(\Omega, x_0, r, R))$$

where

$$w(\Omega, x_0, r, R) = \int_r^R \left(\frac{\mathrm{cap}_p(\Omega^c \cap B_{x_0,t}, B_{x_0,2t})}{\mathrm{cap}_p(B_{x_0,t}, B_{x_0,2t})} \right)^{q-1} \frac{dt}{t},$$
$$osc(h, \Omega) = |\sup_\Omega h(x) - \inf_\Omega h(x)|,$$

and c depends only on the dimension of the space.

Lemma 4.6.6 *Suppose that Ω belongs to $\mathcal{O}_{c,r}(D)$. If $\theta \in H^1(\Omega) \cap C(\Omega)$, and if h is an harmonic function in Ω with $h - \theta \in H_0^1(\Omega)$, then*

$$\lim_{x \to x_0} h(x) = \theta(x_0)$$

for any $x_0 \in \partial\Omega$.

The main continuity result can be expressed as follows:

Theorem 4.6.7 *Let $(\Omega_n)_{n \in \mathbb{N}}$ be a sequence in $\mathcal{O}_{c,r}(D)$, which converges in the H^c topology to an open set Ω. Then Ω_n γ-converges to Ω.*

Proof Let us fix $f \equiv 1$; it will be sufficient to prove the continuity for a subsequence of $\{\Omega_n\}_{n \in \mathbb{N}}$. From Proposition 4.6.2 there exists a subsequence of $\{\Omega_n\}_{n \in \mathbb{N}}$, which we still denote $\{\Omega_n\}_{n \in \mathbb{N}}$, such that $u_{\Omega_n} \rightharpoonup u$ weakly in $H_0^1(D)$, and u satisfies the equation on Ω. We prove that $u \in H_0^1(\Omega)$, which will imply that $u = u_\Omega$. For that it is sufficient to prove $u = 0$ q.e. on $D \setminus \Omega$ where u is a quasi-continuous representative.

From the Banach-Saks theorem there exists a sequence of averages:

$$\psi_n = \sum_{k=n}^{N_n} \alpha_k^n u_{\Omega_n}$$

with

$$0 \le \alpha_k^n \le 1, \quad \sum_{k=n}^{N_n} \alpha_k^n = 1$$

such that

$$\psi_n \xrightarrow{H_0^1(D)} u$$

From the strong convergence of ψ_n to u in $H_0^1(D)$, we have that :

$$\psi_n(x) \longrightarrow u(x) \quad q.e. \quad on \ D$$

for a subsequence of $\{\psi_n\}$ which we still denote by $\{\psi_n\}$.

Let G_0 be the set of zero capacity on which $\psi_n(x)$ does not converge to $u(x)$. Let $x \in D \setminus (\Omega \cup G_0)$, and $\varepsilon > 0$ arbitrary. We prove that $| u(x) | < \varepsilon$. Indeed, we have:

$$|u(x)| \le |u(x) - \psi_n(x)| + |\psi_n(x)|.$$

We consider $n > N_{\varepsilon,x}$ such that

$$|u(x) - \psi_n(x)| < \frac{\varepsilon}{2}$$

Let us consider u_B the solution of (4.32) on a large ball B containing D (with $f \equiv 1$). From the smoothness of B and the regularity of f we have that u_B is continuous on $\overline{B}$. Subtracting the corresponding equations, we obtain:

$$(4.35) \qquad \Delta(u_B - u_{\Omega_n}) = 0 \ in \ \Omega_n$$

So $u_B - u_{\Omega_n}$ is harmonic in Ω_n and continuous on Ω_n. We use Lemma 4.6.6 in the following way: θ is the restriction to Ω_n of the function u_B and

$h_n = u_B\big|_{\Omega_n} - u_{\Omega_n}$. Now $v - \theta = u_{\Omega_n}$ which belongs to $H_0^1(\Omega_n)$. From the continuity of u_B we obtain that the continuous extension of $u_B\big|_{\Omega_n} - u_{\Omega_n}$ is equal to u_B on the boundary of Ω_n, and so the extension of u_{Ω_n} to the boundary is zero. Using Lemma 4.6.5, one obtains (see [136]) that if h_n is Lipschitz on $\partial\Omega_n$ then it is Hölderian on all Ω_n. Since u_B is Lipschitz on B and is equal to h_n on $\partial\Omega_n$ we have

$$\forall x, y \in \partial\Omega_n \quad |h_n(x) - h_n(y)| = |u_B(x) - u_B(y)| \le M|x - y|.$$

There exists two constants $\delta_1 > 0$ and a M_1 given by lemma 4.6.5 (see [136] for more details) which depend only on c, r, *diam* B and the dimensions of the space N such that

$$|h_n(x) - h_n(y)| \le M_1|x - y|^{\delta_1} \quad \forall x, y \in \Omega_n.$$

This inequality holds obviously in B (changing the constant M_1 if necessary), since h_n is equal to u_B outside Ω_n. Therefore for every $x, y \in B$ we have:

$$|u_{\Omega_n}(x) - u_{\Omega_n}(y)| \le |h_n(x) - h_n(y)| + |u_B(x) - u_B(y)| \le$$

$$\le M_1|x - y|^{\delta_1} + M|x - y| \le M_2|x - y|^{\delta_2}$$

Let us choose $R > 0$, such that $M_2 R^{\delta_2} < \varepsilon/2$. From the H^c convergence of Ω_n to Ω there exists $n_R \in \mathbb{N}$, such that for every $n \ge n_R$ we have $(B\backslash\Omega_n) \cap B_{x,R} \ne \emptyset$. Let us take $x_n \in (B \backslash \Omega_n) \cap B_{x,R}$. We have:

$$|u_{\Omega_n}(x)| = |u_{\Omega_n}(x) - u_{\Omega_n}(x_n)| \le M_2|x - x_n|^{\delta_2} \le M_2 R^{\delta_2} \le \frac{\varepsilon}{2}$$

because $u_{\Omega_n}(x_n) = 0$. Hence

$$|\psi_n(x)| = |\sum_{k=n}^{N_n} \alpha_k^n u_{\Omega_n}(x)| \le \sum_{k=n}^{N_n} \alpha_k^n \frac{\varepsilon}{2} = \frac{\varepsilon}{2}, \qquad \forall n > n_R.$$

Finally we obtain $|u(x)| \le \varepsilon$. Since ε was taken arbitrarily we have $u(x) = 0$ q.e. on $B \backslash \Omega$, which implies that $u = u_\Omega$. The strong convergence of u_{Ω_n} to u_Ω is now immediate, from the convergence of the norms of u_{Ω_n} to the norm of u_Ω.

∎

4.7 Continuity under topological constraints: Šverák's result

Let us denote by

$$\mathcal{O}_l(D) = \{\Omega \subseteq D | \sharp\Omega^c \leq l\}$$

the family of open subsets of D whose complementaries have at most l connected components. By $\sharp$ we denote the number of connected components.

A consequence of Theorem 4.6.7 is the following result due to Šverák [177].

Theorem 4.7.1 *Let $N=2$. If $\Omega_n \in \mathcal{O}_l(D)$ and $\Omega_n \xrightarrow{H^c} \Omega$ then Ω_n γ-converges to Ω.*

Proof Let us fix $f \equiv 1$. Since the solution of equation (4.32) is unique, it is sufficient to prove the continuity result for a subsequence.

There exists a subsequence of $\{\Omega_n\}_{n\in\mathbb{N}}$ still denoted $\{\Omega_n\}_{n\in\mathbb{N}}$, such that $u_{\Omega_n} \rightharpoonup u$ weakly in $H_0^1(D)$, and by Proposition 4.6.2 u satisfies the equation on Ω. To obtain that $u \in H_0^1(\Omega)$ we prove that $u = 0$ q.e. on Ω^c.

In general, one cannot find $c, r > 0$ such that $\{\Omega_n\}_{n\in\mathbb{N}} \subseteq \mathcal{O}_{c,r}(D)$, therefore a direct application of Theorem 4.6.7 is not possible. Let

$$\overline{B} \setminus \Omega_n = K_1^n \cup ... \cup K_l^n$$

be the decomposition of $\overline{B} \setminus \Omega_n$ in l connected components, which are compact and disjoint, maybe empty. Following Proposition 4.6.1 there exists a subsequence $(K_1^{k_n^1})_{k_n^1}$ of $(K_1^n)_n$ such that

$$K_1^{k_n^1} \xrightarrow{H} K_1$$

By the same argument we can extract a subsequence of (k_n^1) which we denote (k_n^2) such that:

$$K_2^{k_n^2} \xrightarrow{H} K_2$$

Finally, continuing this procedure, we obtain a subsequence of $\{\Omega_n\}_{n\in\mathbb{N}}$ (still denoted using the same indices) such that:

$$K_j^n \xrightarrow{H} K_j \ \forall j = 1, ..., l.$$

Obviously $\Omega = D \setminus (K_1 \cup ... \cup K_l)$. Since K_j^n are connected, we have that K_j is connected. There are now three possibilities. Either K_j is the empty set, or it is a point, or it contains at least two points; in the latter case any connected open set which contains K_j, also contains a continuous curve which links the two points. If $K_j = \emptyset$ we ignore K_j and $(K_j^n)_n$ which are also empty (for n large enough). If K_j is a point then it has zero capacity. In that case we define the new sets $\Omega_n^+ = K_1^n \cup K_2^n \cup ... \cup K_{j-1}^n \cup K_{j+1}^n \cup ...$. We continue this procedure for all $j = 1, ..., l$ and obtain that:

$$\Omega_n^+ \xrightarrow{H^c} \Omega^+.$$

Of course $u_{\Omega^+} = u_\Omega$ because the difference between Ω^+ and Ω has zero capacity (it consists only of a finite number of points).

Let us prove that there exist $c, r > 0$ such that $(\Omega_n^+)_{n \in \mathbb{N}} \subseteq \mathcal{O}_{c,r}$. There exists $\delta > 0$ such that $diam(K_i) \geq \delta$ for all remaining indices $i \in \{1, ..., l\}$. For n large enough, we get that $diam(K_i^n) \geq \delta/2$.

Let us set

$$c = \frac{cap([0, 1] \times \{0\}, B_{0,2})}{cap(\overline{B}_{0,1}, B_{0,2})} > 0.$$

Then we have that for n large enough $\Omega_n^+ \in \mathcal{O}_{c,r}$ with c previously defined and $r = \delta/4$. In order to prove that for every $x \in \partial \Omega_n^+$ and $t \in (0, r)$ we have

$$\frac{cap((\Omega_n^+)^c \cap B_{x,t}, B_{x,2t})}{cap(B_{x,t}, B_{x,2t})} \geq c,$$

we simply remark that for every $x \in K_i^n$ and $t \in (0, r)$

$$cap(K_i^n \cap B_{x,t}, B_{x,2t}) \geq cap([x, y], B_{x,2t}).$$

Here y is a point belonging on $\partial B_{x,t}$. This inequality follows straight forwardly by Steiner symmetrization. For details, we refer to [46, 32] and Section 6.2.

Using Theorem 4.6.7 we get that

$$u_{\Omega_n^+} \longrightarrow u_{\Omega^+} \text{ strongly in } H_0^1(D).$$

From the maximum principle we have $u_{\Omega_n^+} \geq u_{\Omega_n} \geq 0$, hence $u_{\Omega^+} \geq u \geq 0$, hence $u \in H_0^1(\Omega)$.   ∎

Remark 4.7.2 In three or more dimensions a curve has zero capacity, hence an analogous of Šverák's result can not be obtained for the Laplace operator. In [46] a Šverák type result is proved for the p-Laplacian for $p \in (N-1, N]$. For $p > N$ a trivial shape continuity result holds in the H^c-topology, since a point has strictly positive p-capacity.

Instead of the Laplace operator, in equation (4.32) one could consider an elliptic operator of the form $u \mapsto -div(A(x)\nabla u) + a(x)u$, where $A \in L^\infty(D, \mathbb{R}^{N \times N})$ is such that $\alpha I_d \leq A \leq \beta I_d$ and $a \in L^\infty(D, \mathbb{R}^+)$. Theorem 4.6.7 remains true with the same hypotheses; this is a consequence of the Mosco convergence of the Sobolev spaces. Non homogeneous boundary conditions can obviously be reduced to homogeneous boundary conditions by changing the right hand side of the equation.

The nonlinear case was treated in [46]. There are considered monotone operators equivalent to the p-Laplacian. All shape continuity results hold in similar classes of domains: convex, uniform cone, flat cone, p-capacity density condition, p-uniform Wiener criterion. Several of those results can also be found in [186]. In the next section we give necessary and sufficient conditions for the shape continuity of the solution of the p-Laplacian.

The case of elliptic systems, as the elasticity equations, is treated in [45]; the Stokes equation is discussed in [177]. New difficulties appear when dealing with the convergence in the sense of Mosco of free divergence spaces, mainly because for nonsmooth open sets Ω

$$\{u \in [H_0^1(\Omega)]^N : \text{div } u = 0\} \neq cl_{[H_0^1(\Omega)]^N}\{u \in [C_0^\infty(\Omega)]^N : \text{div } u = 0\}.$$

When studying the shape continuity of evolution equations, it is common to try to prove the following type of result: if shape continuity holds for the corresponding stationary equation, prove the shape continuity for the evolution equation. For the heat equation we refer to the book of Attouch [15], while for more general (degenerate) parabolic problems we refer to [174]. Hyperbolic equations were discussed by Toader in [179].

4.8 Necessary and sufficient conditions for the γ-convergence

In order to familiarize the reader the nonlinear framework, in this section we discuss the necessary and sufficient conditions for the γ_p convergence in terms of the convergence of the local capacities.

Let $(\Omega_n)_{n\in\mathbb{N}}$ be a sequence of open subsets of a bounded design region D, and $1 < p < \infty$. Since the role of the design region is not important in the sequel, we can replace D by a large ball denoted by B such that $D \subseteq B$. Let us fix $f \in W^{-1,q}(B)$ and $g \in W_0^{1,p}(B)$. The purpose of this section is to give necessary and sufficient conditions for the γ_p-continuity, i.e. the continuity of the mapping:

$$(4.36) \qquad n \to u_{\Omega_n,f,g} \quad \text{where} \quad \begin{cases} -\Delta_p u_{\Omega_n,f,g} & = \ f \ \text{in} \ \Omega_n \\ u_{\Omega_n,f,g} & = \ g \ \text{on} \ \partial\Omega_n \end{cases}$$

only in terms of the geometric behavior of the moving domain.

We consider the sequence of the associated Sobolev spaces $(W_0^{1,p}(\Omega_n))_{n\in\mathbb{N}}$, and study the weak upper and strong lower limits in the sense of Kuratowski of this sequence in terms of the behavior of the local capacity of the complementaries intersected with open and closed balls. We refer the reader to [35] for further details.

We write

$$(4.37) \qquad \begin{aligned} & s - \liminf_{n\to\infty} W_0^{1,p}(\Omega_n) \\ & = \{u \in W_0^{1,p}(B) | \exists u_n \in W_0^{1,p}(\Omega_n), \ \text{and} \ u_n \overset{W_0^{1,p}(B)}{\to} u\} \end{aligned}$$

$$(4.38) \qquad \begin{aligned} & w - \limsup_{n\to\infty} W_0^{1,p}(\Omega_n) \\ & = \{u \in W_0^{1,p}(B) | \exists u_{n_k} \in W_0^{1,p}(\Omega_{n_k}), \ \text{and} \ u_{n_k} \overset{W_0^{1,p}(B)}{\rightharpoonup} u\}. \end{aligned}$$

In general

$$s - \liminf_{n\to\infty} W_0^{1,p}(\Omega_n) \subseteq w - \limsup_{n\to\infty} W_0^{1,p}(\Omega_n).$$

Given an open set $\Omega \subseteq B$ we establish necessary and sufficient conditions for each of the following inclusions

$$(4.39) \qquad W_0^{1,p}(\Omega) \subseteq s - \liminf_{n\to\infty} W_0^{1,p}(\Omega_n)$$

and

$$(4.40) \qquad w - \limsup_{n\to\infty} W_0^{1,p}(\Omega_n) \subseteq W_0^{1,p}(\Omega).$$

When these two limits coincide, we obtain a characterization for the γ_p-convergence. The Mosco convergence of these spaces was studied by Dal Maso [91], [92] and by Dal Maso-Defranceschi [93]. Using the frame of the relaxation theory, they proved that Ω_n γ_p-converges to Ω if and only if there exists a family of sets $\mathcal{A} \subseteq \mathcal{P}(B)$ which is rich or dense (see the exact definitions in [92]) in $\mathcal{P}(B)$ such that

$$(4.41) \qquad \mathrm{cap}_p(\Omega^c \cap X, B) = \lim_{n \to \infty} \mathrm{cap}_p(\Omega_n^c \cap X, B) \quad \forall X \in \mathcal{A}.$$

By considering for example monotone sequences of the family $\mathcal{P}(B)$, in [92] it is shown that the convergence in the sense of Mosco is still equivalent to the following two relations which have to be satisfied for all p-quasi-open sets $A \subseteq B$ and p quasi-compacts sets $F \subseteq A \subseteq B$.

$$(4.42) \qquad \mathrm{cap}_p(\Omega^c \cap A, B) \geq \limsup_{n \to \infty} \mathrm{cap}_p(\Omega_n^c \cap F, B).$$

and

$$(4.43) \qquad \mathrm{cap}_p(\Omega^c \cap A, B) \leq \liminf_{n \to \infty} \mathrm{cap}_p(\Omega_n^c \cap A, B)$$

In the sequel, we prove that (4.39) is equivalent to a simpler version of (4.42) (where capacity is calculated by intersection with closed balls) and (4.40) is equivalent to a simpler version of (4.43) (where capacity is calculated on open balls).

We study independently each one of the Kuratowski limits. For a given sequence of open sets, we find a sequence of subdomains, respectively super domains, which γ_p-converge to the target domain, and for which relations (4.42)-(4.43) hold. For the necessity, we use the results of Dal Maso, replacing the p quasi-open sets by open balls and the p quasi-closed sets by closed balls. For the sufficiency, the study of the strong lower limit is made directly, being given a constructive way to find the strong approximation for each element of $W_0^{1,p}(\Omega)$. The most difficult part concerns the weak upper limit, which is related to the study of the behavior of particular solutions of equation (4.36), for $f \equiv 0$. For simplicity let us denote in the sequel $v_{\Omega_n,g} = u_{\Omega_n,0,g}$, and when no ambiguity occurs, $v_n = v_{\Omega_n,g}$.

Study of the strong lower limit. We prove the following

Theorem 4.8.1 *Let $(\Omega_n)_{n \in \mathbb{N}}, \Omega$ be open subsets of B. Then*

$$W_0^{1,p}(\Omega) \subseteq s - \liminf_{n \to \infty} W_0^{1,p}(\Omega_n)$$

if and only if for every $x \in \mathbb{R}^N$ and $\delta > 0$

$$(4.44) \qquad \operatorname{cap}_p(\Omega^c \cap \overline{B}_{x,\delta}, B_{x,2\delta}) \geq \limsup_{n \to \infty} \operatorname{cap}_p(\Omega_n^c \cap \overline{B}_{x,\delta}, B_{x,2\delta}).$$

In a first step we prove that if relation (4.39) holds then we can find a sequence of subsets of Ω_n which γ_p-converges to Ω. This is sufficient to apply the results of [92] and to use the behavior of the capacity on decreasing sequence of compacts. In relation (4.42) one should replace F by a closed ball, and choose for A a decreasing sequence of open balls containing F. For the sufficiency, we give a constructive way to approach any element of $\mathcal{D}(\Omega)$ by a sequence of elements of $W^{1,p}(\Omega_n)$.

Proof [of Theorem 4.8.1] ($\Rightarrow$)   Remark that if (4.39) holds, then $\Omega_n \cap \Omega$ γ_p-converges to Ω. We verify the two conditions (4.39)-(4.40). For (4.40) consider $u_n \in W_0^{1,p}(\Omega_n \cap \Omega)$ and $u_n \rightharpoonup u$ weakly in $W_0^{1,p}(B)$. Obviously we have $u \in W_0^{1,p}(\Omega)$. To prove relation (4.39), let us consider $u \in W_0^{1,p}(\Omega)$. Without loosing the generality one can suppose that $u \geq 0$ (if not we decompose $u = u^+ - u^-$). By hypothesis, there exists a sequence (which can be taken positive) $u_n \in W_0^{1,p}(\Omega_n)$ such that $u_n \to u$ strongly in $W_0^{1,p}(B)$. Then $\min\{u_n, u\} \in W_0^{1,p}(\Omega_n \cap \Omega)$ and $\min\{u_n, u\} \to u$ strongly in $W_0^{1,p}(B)$.

In order to localize the moving domain in a ball, by a similar argument we get that $\Omega_n \cap \Omega \cap B_{x,2r}$ γ_p-converges to $\Omega \cap B_{x,2r}$. Therefore one can apply (4.42) for $F = \overline{B}_{x,r}$, $A = B_{x,r+\varepsilon}$, and get

$$\operatorname{cap}_p(\Omega^c \cap B_{x,r+\varepsilon}, B_{x,2r}) \geq \limsup_{n \to \infty} \operatorname{cap}_p((\Omega_n \cap \Omega)^c \cap \overline{B}_{x,r}, B_{x,2r}).$$

Since

$$\operatorname{cap}_p(\Omega^c \cap \overline{B}_{x,r+\varepsilon}, B_{x,2r}) \geq \operatorname{cap}_p(\Omega^c \cap B_{x,r+\varepsilon}, B_{x,2r}) \geq \operatorname{cap}_p(\Omega^c \cap \overline{B}_{x,r}, B_{x,2r})$$

using the continuity of the capacity on decreasing sequences of compacts we get making $\varepsilon \to 0$,

$$\operatorname{cap}_p(\Omega^c \cap \overline{B}_{x,r}, B_{x,2r}) \geq \limsup_{n \to \infty} \operatorname{cap}_p((\Omega_n \cap \Omega)^c \cap \overline{B}_{x,r}, B_{x,2r}).$$

Since $\mathrm{cap}_p((\Omega_n \cap \Omega)^c \cap \overline{B}_{x,r}, B_{x,2r}) \geq \mathrm{cap}_p(\Omega_n^c \cap \overline{B}_{x,r}, B_{x,2r})$ we obtain (4.44).

($\Leftarrow$) Let us consider $u \in C_c^\infty(\Omega)$, *supp* $u = K \subset\subset \Omega$ and $\varepsilon = d(K, \partial\Omega)$. There exist a finite family of k balls centered in points of K and of radius less than ε such that $K \subseteq \bigcup_{r=1,k} B_{x_r, \varepsilon/2}$. We also have

$$\mathrm{cap}_p(\Omega^c \cap \overline{B}_{x_r, \varepsilon/2}, B_{x_r, \varepsilon}) = 0$$

and thus we get from (4.44)

$$\lim_{n\to\infty} \mathrm{cap}_p(\Omega_n^c \cap \overline{B}_{x_r, \varepsilon/2}, B_{x_r, \varepsilon}) = 0.$$

Hence the capacity of $\Omega_n^c \cap \overline{B}_{x_r, \varepsilon/2}$ vanishes as $n \to \infty$ and we consider for $r = 1, ..., k$ smooth functions $\psi_n^r \in \mathcal{D}(B_{x,\varepsilon})$ equal to 1 on $\Omega_n^c \cap \overline{B}_{x_r, \varepsilon/2}$ and which approximate respectively the capacity of this set, namely

$$\|\psi_n^r\|_{W_0^{1,p}(B)} \leq \mathrm{cap}_p(\Omega_n^c \cap \overline{B}_{x_r, \frac{\varepsilon}{2}}, B_{x_r, \varepsilon}) + \frac{1}{n}.$$

Moreover the functions ψ_n^r can be chosen such that $0 \leq \psi_n^r \leq 1$. Therefore, since k is fixed, the sequence of functions defined by

$$u_n = u \prod_{r=1}^{k} (1 - \psi_n^r)$$

has the property

$$u_n \in W_0^{1,p}(\Omega_n) \quad and \quad u_n \xrightarrow{W_0^{1,p}(B)} u,$$

so (4.39) is satisfied.

The passage from $\mathcal{D}(\Omega)$ to $W_0^{1,p}(\Omega)$ is made using the density $\overline{\mathcal{D}(\Omega)} = W_0^{1,p}(\Omega)$ and a usual diagonal procedure for extracting a convergent sequence. ∎

If one replaces in the theorem above the sequence $(\Omega_n)_{n\in\mathbb{N}}$ of open by p quasi-open sets, the only technical point which does not work concerns the necessity. The construction of the approximating sequence fails because the dense family in $W_0^{1,p}(\Omega)$, which replaces $\mathcal{D}(\Omega)$ for a p quasi-open set Ω, does not have the same properties. In fact, even for a p quasi-open set Ω, there exists a dense family of functions having compact support in Ω, but their support cannot be covered by balls which do not intersect the complement of Ω (see [140]).

Study of the weak upper limit. We prove the following

Theorem 4.8.2 *Let $(\Omega_n)_{n\in\mathbb{N}}, \Omega$ be open subsets of B. Then*

$$w - \limsup_{n\to\infty} W_0^{1,p}(\Omega_n) \subseteq W_0^{1,p}(\Omega)$$

if and only if $\forall x \in \mathbb{R}^N, \forall \delta > 0$

$$(4.45) \qquad \operatorname{cap}_p(\Omega^c \cap B_{x,\delta}, B_{x,2\delta}) \le \liminf_{n\to\infty} \operatorname{cap}_p(\Omega_n^c \cap B_{x,\delta}, B_{x,2\delta}).$$

Proof ($\Rightarrow$) The necessity does not follow as directly as in Theorem 4.8.1; the idea is still the same: find a sequence of open sets which contain Ω_n and which γ_p-converges to Ω. For this purpose we give the next lemma which is the nonlinear version of a result obtained in [64].

Lemma 4.8.3 *Let be given a sequence of open sets Ω_n for which*

$$u_{\Omega_n,1,0} \overset{W_0^{1,p}(B)}{\rightharpoonup} w$$

and $w \in W_0^{1,p}(\Omega)$. There exists a subsequence (still denoted using the same indices) and a sequence of open sets $G_n \subseteq B$ with $\Omega_n \subseteq G_n$ and G_n γ_p-converges to Ω.

Proof Denote by $A \supseteq \{w > 0\}$ a p quasi-open set containing the region where the limit function is positive. Following [100] we have $w \le u_{A,1,0}$. For each $\varepsilon > 0$ we define, as is done in [64] for $p = 2$, the following p quasi-open set $A^\varepsilon = \{u_{A,1,0} > \varepsilon\}$. For a subsequence, still denoted using the same indices, we can suppose that

$$u_{\Omega_n \cup A^\varepsilon,1,0} \overset{W_0^{1,p}(B)}{\rightharpoonup} w^\varepsilon$$

and by the comparison principle we have that $w^\varepsilon \ge u_{A^\varepsilon,1,0}$. But $w^\varepsilon \in W_0^{1,p}(A)$ from Lemma 4.3.11 applied to $\{\Omega_n \cup A^\varepsilon\}$. Indeed, defining $v^\varepsilon = 1 - \frac{1}{\varepsilon}\min\{u_{A,1,0}, \varepsilon\}$ we get $0 \le v^\varepsilon \le 1$ and $v^\varepsilon = 0$ on A^ε, $v^\varepsilon = 1$ on $B \setminus A$. Taking $u_n = \min\{v^\varepsilon, u_{\Omega_n \cup A^\varepsilon,1,0}\}$ we get $u_n = 0$ on $A^\varepsilon \cup (B \setminus (\Omega_n \cup A^\varepsilon))$, particularly on $B \setminus \Omega_n$. Moreover $u_n \rightharpoonup \min\{v^\varepsilon, w^\varepsilon\}$ weakly in $W_0^{1,p}(B)$ and hence $\min\{v^\varepsilon, w^\varepsilon\}$ vanishes p-q.e. on $\{w = 0\}$. Since $v^\varepsilon = 1$ on $B \setminus A$ we get that $w^\varepsilon = 0$ p q.e. on $B \setminus A$.

Using [100, Theorem 5.1], from the fact that $-\Delta_p u_{\Omega_n \cup A^\varepsilon, 1, 0} \leq 1$ in B we get $-\Delta_p w^\varepsilon \leq 1$ and hence $w^\varepsilon \leq u_{A, 1, 0}$. Finally $u_{A^\varepsilon, 1, 0} \leq w^\varepsilon \leq u_{A, 1, 0}$, and by a diagonal extraction procedure we get that

$$u_{\Omega_n \cup A^{\varepsilon_n}, 1, 0} \overset{W_0^{1,p}(B)}{\rightharpoonup} u_{A, 1, 0}$$

Taking Ω in the place of the set A, we conclude the proof. ∎

Proof [of Theorem 4.8.2, continuation] Using Lemma 4.8.3, we consider the sequence $G_n \cap B_{x, 2r}$ which γ_p-converges to $\Omega \cap B_{x, 2r}$. Taking in (4.43) $A = B_{x, r}$ we get

$$\mathrm{cap}_p(\Omega^c \cap B_{x, r}, B_{x, 2r}) \leq \liminf_{n \to \infty} \mathrm{cap}_p(G_n^c \cap B_{x, r}, B_{x, 2r}).$$

Since $\mathrm{cap}_p(G_n^c \cap B_{x, r}, B_{x, 2r}) \leq \mathrm{cap}_p(\Omega_n^c \cap B_{x, r}, B_{x, 2r})$ inequality (4.45) is proved.

($\Leftarrow$) In order to reduce the study of arbitrary weak convergent sequences to particular sequences of solutions of equation (4.36) for $f \equiv 0$, we give the following lemma (recall the notation $v_{\Omega, g} = u_{\Omega, 0, g}$).

Lemma 4.8.4 *Let $(\Omega_n)_{n \in \mathbb{N}}$ be a sequence of open subsets of B, and Ω an open set such that for every $g \in \mathcal{D}(B)$ and for every $W_0^{1,p}(B)$-weak limit v of a sequence $(v_{\Omega_{n_k}, g})$ we have $v - g \in W_0^{1,p}(\Omega)$. Then relation (4.40) holds.*

Proof Remark first that if the conclusion holds for any $g \in \mathcal{D}(B)$ then it holds for any $g \in W_0^{1,p}(B)$. Indeed, consider some $g \in W_0^{1,p}(B)$ and (with a renotation of the indices) suppose that $v_{\Omega_n, g} \rightharpoonup v_g$ weakly in $W_0^{1,p}(B)$. There exists a sequence $g_k \in \mathcal{D}(B)$ such that $g_k \to g$ strongly in $W_0^{1,p}(B)$. Following [100], one can find a uniform bound for v_{Ω_n, g_k}, hence there exists a constant β such that

$$\|v_{\Omega_n, g_k} - v_{\Omega_n, g}\|_{W^{1,p}(B)} \leq \beta \|g - g_k\|_{W_0^{1,p}(B)}$$

for all $k \in \mathbb{N}$. Then $\|v - v_k\|_{W^{1,p}(B)} \leq \beta \|g - g_k\|_{W_0^{1,p}(B)}$ and if $v_k - g \in W_0^{1,p}(\Omega)$ we conclude that $v - g \in W_0^{1,p}(\Omega)$.

For proving (4.40) we consider a sequence $u_n \rightharpoonup u$ weakly in $W_0^{1,p}(B)$, with $u_n \in W_0^{1,p}(\Omega_n)$. Following [100], for a subsequence still denoted using the same indices, Ω_n γ_p-converges to $(\mathbb{R}^N, \mu)$, μ being the measure defined by (4.17). Then we have $u \in W_0^{1,p}(B) \cap L_\mu^p(B)$. To prove that $u \in W_0^{1,p}(\Omega)$ it suffices to verify that

$$(4.46) \qquad W_0^{1,p}(B) \cap L_\mu^p(B) \subseteq W_0^{1,p}(\Omega)$$

Consider some $g^* \in W_0^{1,p}(B)$. For a subsequence still denoted using the same indices, we have $v_{\Omega_n, g^*} \rightharpoonup v$ weakly in $W_0^{1,p}(B)$ and v satisfies the equation

$$(4.47) \qquad \begin{cases} -\Delta_p v + \mu|v - g^*|^{p-2}(v - g^*) = 0 \ \ in \ W_0^{1,p}(B) \cap L_\mu^p(B) \\ \qquad\qquad\qquad\qquad\qquad v - g^* \in W_0^{1,p}(B) \cap L_\mu^p(B). \end{cases}$$

Hence on one side we obtain that $v - g^* \in W_0^{1,p}(B) \cap L_\mu^p(B)$, and on the other side the hypothesis we assumed gives $v - g^* \in W_0^{1,p}(\Omega)$. To obtain the conclusion it is sufficient to prove that the family of functions written in the form $v - g^*$ with the properties above, is dense in $W_0^{1,p}(B) \cap L_\mu^p(B)$. This will insure inclusion (4.46). Set $v - g^* = z$. We have

$$(4.48) \qquad \begin{cases} -\Delta_p(z + g^*) + \mu|z|^{p-2}z = 0 \ \ in \ W_0^{1,p}(B) \cap L_\mu^p(B) \\ \qquad\qquad\qquad\qquad z \in W_0^{1,p}(B) \cap L_\mu^p(B). \end{cases}$$

The family $\{\varphi w\}_{\varphi \in \mathcal{D}(B)}$, where $\varphi \in C_c^\infty(B)$, and $w = u_{A_\mu, 1, 0}$, being A_μ the regular set of the measure μ (i.e. the union of all finely open sets of finite μ-measure), is dense in $W_0^{1,p}(B) \cap L_\mu^p(B)$. Fix now some $z = \varphi w$. For this z we can prove the existence of some $g \in W_0^{1,p}(B)$ such that

$$-\Delta_p(z + g) + \mu|z|^{p-2}z = 0.$$

Indeed, the existence of such a g is trivial if $\mu|z|^{p-2}z \in W^{-1,q}(B)$. This follows immediately from the particular structure of z. Consider $\theta \in \mathcal{D}(B)$. Then

$$< \mu|z|^{p-2}z, \theta >_{W^{-1,q}(B) \times W_0^{1,p}(B)} = \int_{\{w>0\}} \theta|z|^{p-2}z \, d\mu$$

$$= \int_{\{w>0\}} \theta|z|^{p-2}z \frac{d\nu}{w^{p-1}} = < \theta|\varphi|^{p-2}\varphi, 1 + \Delta_p w >_{W^{-1,q}(B) \times W_0^{1,p}(B)} .$$

Since $\Delta_p w \in W^{-1,q}(B)$ we get $\mu|z|^{p-2}z \in W^{-1,q}(B)$ and this concludes the proof. ∎

Proof [of Theorem 4.8.2, continuation] Following Lemma 4.8.4, it suffices to study the behavior of $v_{\Omega_n,g}$. Therefore, let us consider some $g \in \mathcal{D}(B)$ and $v_{\Omega_n,g} \rightharpoonup v$ weakly in $W_0^{1,p}(B)$.

From [133] it is sufficient to prove $v = g$ q.e on Ω^c. In fact we use an estimation of the oscillation of $v_{\Omega_n,g}$ near the boundaries. The lower semi-continuity of the capacity will insure a common behavior. There exist convex combinations

$$\phi_n = \sum_{k=n}^{N_n} \alpha_k^n v_k \overset{W_0^{1,p}(B)}{\longrightarrow} v$$

and for a subsequence (still denoted using the same indices) it converges p q.e. (for p quasi-continuous representatives). Let us consider $x \in B \setminus E$ the set of points where the convergence is pointwise, with $\mathrm{cap}_p(E) = 0$. So, for $x \in B \setminus E$, for every $\varepsilon > 0$ we have

$$|v(x) - g(x)| \le |v(x) - \phi_n(x)| + |\phi_n(x) - g(x)|$$

and for n large enough we also have $|v(x) - \phi_n(x)| < \varepsilon/2$. We must prove that we have $|\phi_n(x)| < \varepsilon/2$ or even better, that $|v_n(x) - g(x)| < \varepsilon/2$ for n large enough.

One can apply directly Lemma 4.6.5 for $h = v_{\Omega,g}$ and $\theta = g$ (supposed to be α-Hölderian) and get for a point $x_0 \in \partial\Omega$, for all $x, y \in \Omega \cap B_{x_0,r}$

$$(4.49) \qquad |v_{\Omega,g}(x) - v_{\Omega,g}(y)| \le c_g M(4R)^\alpha + 2M \exp(-cw(\Omega, x_0, r, R)).$$

If y is a regular point belonging to the set $\partial\Omega \cap B_{x_0,r}$ then $v_{\Omega,g}(y) = g(y)$ (see also [2, 3]) and one can derive

$$(4.50) \qquad |v_{\Omega,g}(x) - g(x)| \le 2c_g M(4R)^\alpha + 2M \exp(-cw(\Omega, x_0, r, R))$$

for all $x \in \Omega \cap B_{x_0,r}$.

This last inequality proves that if one can handle the behavior of the function $n \to w(\Omega_n, x_0, r, R)$, then the oscillations of $v_{\Omega_n,g}$ relatively to g are *uniform* in some neighborhoods of x_0.

We use Lemma 4.6.5 and we distinguish between two situations $x \in \partial\Omega$ and $x \in ext(\Omega)$. Let us prove first this assertion for $x \in \partial\Omega$. Since the set of non regular points of the boundary of Ω has zero capacity, without loosing the generality one can suppose that x is regular. Let us give two lemmas.

Lemma 4.8.5 *For all $x \in \mathbb{R}^N$, and for all $0 < r < R$ we have*

$$\liminf_{n \to \infty} w(\Omega_n, x, r, R) \geq w(\Omega, x, r, R).$$

Proof The proof of this lemma is immediate from the lower semi-continuity of the local capacity, and the properties of the Lebesgue integral (Fatou's lemma and (4.45)). ∎

Lemma 4.8.6 *There exists a positive constant $\bar{c}$ depending only on the dimension of the space, such that for all $R > r > 0$, for all $x_1, x_2 \in \mathbb{R}^N$ with $|x_1 - x_2| = \delta \leq r/2$ we have*

$$w(\Omega, x_1, r, R) \geq \bar{c} w(\Omega, x_2, \frac{r}{2}, \frac{R}{2}).$$

Proof We have $\Omega^c \cap B_{x_2, \mu} \subseteq \Omega^c \cap B_{x_1, \varepsilon}$ if $\mu + \delta \leq \varepsilon$. Hence for any $R \geq t \geq r \geq 2\delta$ we have

$$\Omega^c \cap B_{x_2, t/2} \subseteq \Omega^c \cap B_{x_1, t}$$

since $\frac{t}{2} + \delta \leq t$. So we get the inclusion $B_{x_1, 2t} \subseteq B_{x_2, 4t}$ and then we can write

$$\mathrm{cap}_p(\Omega^c \cap B_{x_2, t/2}, B_{x_2, 4t}) \leq \mathrm{cap}_p(\Omega^c \cap B_{x_1, t}, B_{x_1, 2t}).$$

Following [136] there exists a constant ξ depending only on the dimension N, such that

$$\xi \, \mathrm{cap}_p(\Omega^c \cap B_{x_2, \frac{t}{2}}, B_{x_2, t})$$

$$\leq \mathrm{cap}_p(\Omega^c \cap B_{x_2, \frac{t}{2}}, B_{x_2, 4t}) \leq \mathrm{cap}_p(\Omega^c \cap B_{x_1, t}, B_{x_1, 2t}).$$

Hence

$$\xi^{q-1} \int_r^R \left(\frac{\mathrm{cap}_p(\Omega^c \cap B_{x_2, \frac{t}{2}}, B_{x_2, t})}{\mathrm{cap}_p(B_{x_2, t}, B_{x_2, 2t})} \right)^{q-1} \frac{dt}{t}$$

$$\leq \int_r^R \left(\frac{\mathrm{cap}_p(\Omega^c \cap B_{x_1, t}, B_{x_1, 2t})}{\mathrm{cap}_p(B_{x_1, t}, B_{x_1, 2t})} \right)^{q-1} \frac{dt}{t}$$

or, making a change of variables in the first integral, and using the behavior of the capacity on homothetic sets we get

$$\frac{\xi^{q-1}}{2^{(N-2)(q-1)}} \int_{\frac{r}{2}}^{\frac{R}{2}} \left(\frac{\text{cap}_p(\Omega^c \cap B_{x_2,t}, B_{x_2,2t})}{\text{cap}_p(B_{x_2,t}, B_{x_2,2t})} \right)^{q-1} \frac{dt}{t}$$

$$\leq \int_r^R \left(\frac{\text{cap}_p(\Omega^c \cap B_{x_1,t}, B_{x_1,2t})}{\text{cap}_p(B_{x_1,t}, B_{x_1,2t})} \right)^{q-1} \frac{dt}{t}.$$

Setting $(\xi 2^{2-N})^{q-1} = \bar{c}$ we get

$$w(\Omega, x_1, r, R) \geq \bar{c} w(\Omega, x_2, \frac{r}{2}, \frac{R}{2})$$

as soon as $|x_1 - x_2| \leq \frac{r}{2} < \frac{R}{2}$. ∎

Proof [of Theorem 4.8.2, continuation] Let us consider $x \in \partial\Omega$ a regular point, i.e. $\lim_{r \to 0} w(\Omega, x, r, R) = \infty$. We shall fix later $r, R > 0$ such that $w(\Omega, x, r/2, R/2) > M$. The value of M will be also precised. If $|x_n - x| \leq r/2$ we have from the previous lemma

$$w(\Omega_n, x_n, r, R) \geq \bar{c} w(\Omega_n, x, \frac{r}{2}, \frac{R}{2})$$

From Lemma 4.8.5, for n large enough one can write

$$w(\Omega_n, x, \frac{r}{2}, \frac{R}{2}) \geq \frac{1}{2} w(\Omega, x, \frac{r}{2}, \frac{R}{2})$$

which implies

$$w(\Omega_n, x_n, r, R) \geq \frac{\bar{c}M}{2}$$

independently on the choice of x_n with $|x_n - x| \leq \frac{r}{2}$.

If $x \in \Omega_n^c$, then p quasi-everywhere we have $v_n(x) = g(x)$. Let us suppose that $x \in \Omega_n$. Since $\text{cap}_p(\Omega^c \cap B_{x,\delta}, B_{x,2\delta}) > 0$ for any $\delta > 0$ (the point x being regular) we have $\text{cap}_p(\Omega_n^c \cap B_{x,\delta}, B_{x,2\delta}) > 0$ for n large enough. We fix $\delta = r/2$ and consider $x_n \in \partial\Omega_n$ (precisely $x \in \Omega_n$, $x_n \in B_{x,\delta} \cap \partial\Omega_n$) and x_n regular. The existence of such a point follows from the fact that $\text{cap}_p(\partial\Omega_n \cap B_{x,\delta}, B_{x,2\delta}) > 0$ via the following lemma.

Lemma 4.8.7 *Let Ω be an open set such that $\Omega \cap B_{x,\delta} \neq \emptyset$ and $\text{cap}_p(\Omega^c \cap B_{x,\delta}, B_{x,2\delta}) > 0$. Then $\text{cap}_p(\partial\Omega \cap B_{x,\delta}, B_{x,2\delta}) > 0$.*

Proof Let us suppose the contrary, namely that $\text{cap}_p(\partial\Omega \cap B_{x,\delta}, B_{x,2\delta}) = 0$ which implies that $\text{cap}_p(ext\Omega \cap B_{x,\delta}, B_{x,2\delta}) > 0$. Without loosing the generality we consider $x = 0$, $\delta = 1$ and the radial transformation $B_{0,1} \to \mathbb{R}^N$ given by $y \xrightarrow{T} \frac{y}{1-|y|}$. This is a bi-Lipschitz transformation restricted to every closed ball $\overline{B}_{0,\mu}$ with $\mu < 1$. Hence the following Sobolev capacity $\text{cap}_p(\partial T(\Omega) \cap B_{0,\frac{\mu}{1-\mu}})$ vanishes since $\text{cap}_p(\partial\Omega \cap \overline{B}_{0,\mu}) = 0$. Since for an increasing sequence of sets the capacity is continuous, we get $\text{cap}_p(\partial T(\Omega \cap B_{0,1})) = 0$ which from [136] means that $T(\Omega \cap B_{0,1}) \cup extT(\Omega \cap B_{0,1})$ is connected, in contradiction with the fact that the two sets are open and disjoint.   ∎

Proof [of Theorem 4.8.2, continuation] One can then write $v_n(x_n) = g(x_n)$ and using relation (4.49) we get

$$|v_n(x) - g(x)| \leq c_g M (4R)^\alpha + 2M \exp(-cw(\Omega_n, x_n, r, R)).$$

Now we fix $r, R, \overline{M}$ such that

$$R = \frac{1}{4}\log_\alpha \frac{\varepsilon}{4c_g M}, \quad \overline{M} = -\frac{2}{c\overline{c}}ln\frac{\varepsilon}{12M}$$

and $r < R/2$ such that $g(\Omega, x, r/2, R/2) > \overline{M}$. Globally we have

$$c_g M (4R)^\alpha + 2M exp(-c\overline{c}\frac{\overline{M}}{2}) \leq \frac{\varepsilon}{2}$$

and for n large enough, such that $g(\Omega_n, x_n, r, R) \geq \overline{c}\overline{M}/2$ for $|x_n - x| \leq r/2$, also $\text{cap}_p(\Omega_n^c \cap B_{x,r/2}, B_{x,r}) > 0$. Hence $|v_n(x) - g(x)| < \varepsilon/2$ for n large enough, which finally implies $|v(x) - g(x)| \leq \varepsilon$. Since ε was arbitrarily chosen we get $v(x) = g(x)$ for q.e $x \in \partial\Omega$.
   For the case $x \in ext(\Omega)$, the same proof works with $w(\Omega, x, r, R) = \int_r^R \frac{1}{t}dt$.   ∎

In Theorem 4.8.2 one can replace open sets by p quasi-open sets. For the necessity the same proof follows. For the sufficiency, if $(A_n)_{n \in \mathbb{N}}$ and A are p quasi-open subsets of B satisfying (4.45), we construct the open sets Ω_n such that $A_n \subseteq \Omega_n$ and $\text{cap}_p(\Omega_n \setminus A_n) \leq 1/n$, respectively $A \subseteq \Omega_\varepsilon$ and

$\mathrm{cap}_p(\Omega_\varepsilon \setminus A) \leq \varepsilon$. One sees that relation (4.45) still holds for Ω_n and for Ω_ε. Hence we apply Theorem 4.8.2 and get

$$w - \limsup_{n \to \infty} W_0^{1,p}(\Omega_n) \subseteq W_0^{1,p}(\Omega_\varepsilon).$$

Since $W_0^{1,p}(A_n)$ has the same Kuratowski limits as $W_0^{1,p}(\Omega_n)$ we get

$$w - \limsup_{n \to \infty} W_0^{1,p}(A_n) \subseteq W_0^{1,p}(\Omega_\varepsilon).$$

This inclusion holds for any $\varepsilon > 0$, therefore we can replace Ω_ε by A.

Remark 4.8.8 Note that for some fixed $x \in \mathbb{R}^N$, the family of $t \in \mathbb{R}_+$ such that the following strict inequality holds

$$\mathrm{cap}_p(\Omega^c \cap \overline{B}_{x,t}, B_{x,2t}) > \mathrm{cap}_p(\Omega^c \cap B_{x,t}, B_{x,2t})$$

is at most countable (see [35]). Hence, we can state that Ω_n γ_p-converges to Ω if and only if for every $x \in \mathbb{R}^N$ there exists an at most countable family $T_x \subseteq \mathbb{R}_+$ such that for all $t \in \mathbb{R}_+ \setminus T_x$ we have

$$\lim_{n \to \infty} \mathrm{cap}_p(\Omega_n^c \cap B_{x,t}, B_{x,2t}) = \mathrm{cap}_p(\Omega^c \cap B_{x,t}, B_{x,2t}).$$

4.9 Stability in the sense of Keldysh

In a large class of problems, the shape stability question for the solution of the following elliptic equation

(4.51)
$$\begin{cases} -\Delta_p u_\Omega = f \text{ in } \Omega \\ u_\Omega \in W_0^{1,p}(\Omega) \end{cases}$$

has the following formulation: *let $(\Omega_n)_{n \in \mathbb{N}}$ be a perturbation of an open set Ω; the question is whether the solution u_{Ω_n} of equation (4.51) on Ω_n converges in $W_0^{1,p}(B)$ to u_Ω.*

Several papers in the literature deal with the perturbation given by the so called *compact convergence*, which is a sequential topology defined as follows: Ω_n compactly converges to Ω if for every compact $K \subset \Omega \cup \overline{\Omega}^c$ there exists $n_K \in \mathbb{N}$ such that for all $n \geq n_K$, $K \subseteq \Omega_n \cup \overline{\Omega}_n^c$ (see [134], [171]).

Let us remark that the way is defined the topology hides condition (4.39). Shape stability holds for this kind of perturbations (see [134, 171]) as soon as the limit set Ω is p-stable, i.e. for any function $u \in W^{1,p}(\mathbb{R}^N)$ vanishing a.e on $\overline{\Omega}^c$ we have $u \in W_0^{1,p}(\Omega)$. This is a notion introduced in the linear case by Keldysh (see [141], [133]). Roughly speaking, open sets with cracks are not stable. Notice that, for this kind of perturbations, the stability of the solution depends only on Ω. No regularity assumption is made on the converging sequence (Ω_n). This is the main reason for which it is of interest to characterize the p-stable domains.

Based on the results of the previous section, we give a simple proof of the characterization of the p-stable domains.

Proposition 4.9.1 *A bounded open set Ω is p-stable if and only if for every $x \in \mathbb{R}^N, r > 0$ we have*

$$(4.52) \qquad \text{cap}_p(B_{x,r} \setminus \Omega, B_{x,2r}) = \text{cap}_p(B_{x,r} \setminus \overline{\Omega}, B_{x,2r}).$$

Proof The proof is an immediate consequence of the fact that $\Omega_n = \cup_{x \in \Omega} \overset{\bullet}{B}_{x,1/n}$ compactly converges to Ω. If Ω is p-stable, then Ω_n γ_p-converges to Ω and writing (4.45) we get

$$\text{cap}_p(B_{x,r} \setminus \Omega, B_{x,2r}) \le \liminf_{n \to \infty} \text{cap}_p(B_{x,r} \setminus \Omega_n, B_{x,2r})$$

The behavior of the capacity on increasing sequences gives

$$\lim_{n \to \infty} \text{cap}_p(B_{x,r} \setminus \Omega_n, B_{x,2r}) = \text{cap}_p(B_{x,r} \setminus \overline{\Omega}, B_{x,2r}),$$

hence $\text{cap}_p(B_{x,r} \setminus \Omega, B_{x,2r}) \le \text{cap}_p(B_{x,r} \setminus \overline{\Omega}, B_{x,2r})$. The equality follows from the monotonicity of the capacity in the first argument.

Conversely, relation (4.52) yields that Ω_n γ_p-converges to Ω. Then any function $u \in W^{1,p}(\mathbb{R}^N)$ with $u = 0$ a.e. on Ω^c has the property that $u \in W_0^{1,p}(\Omega_n)$, since $u = 0$ p-q.e. on Ω_n^c. The γ_p-convergence of Ω_n to Ω gives $u \in W_0^{1,p}(\Omega)$, hence Ω is p stable. ∎

Chapter 5

Existence of classical solutions

Let $\mathcal{A}$ be a class of admissible open (or, if specified, quasi-open) subsets of the design region D and $F : \mathcal{A} \to [0, +\infty]$ be a functional such that F is γ-lower semicontinuous. Our purpose is to look for the existence of a minimizer for the following problem.

$$(\wp) \qquad\qquad \min\{F(\Omega) : |\Omega| \leq m, \Omega \in \mathcal{A}\}$$

The γ-convergence on the family of *all* open (or quasi-open) subsets of D is not compact if the dimension N is greater than 1; indeed several shape optimization problems of the form $(\wp)$ do not admit any solution, and the introduction of relaxed formulation is needed in order to describe the behavior of minimizing sequences.

Even if in general problem $(\wp)$ does not admit a solution, some particular cases of existence results are available provided that either the cost functional F is *regular* in some sense or the family of admissible domains $\mathcal{A}$ is smaller. This is for example the case when the cost functional F is monotone decreasing with respect to the set inclusion or if we search the minimizer in a class of admissible domains on which we impose some geometrical constraints.

5.1 Existence of optimal domains under geometrical constraints

In the Chapter 4 we proved the continuity in the H^c topology of the solution of (4.32) in several classes of domains. In order to deduce that these classes are γ-compact, it would be sufficient to prove that they are closed in the H^c-topology and that the Lebesgue measure is l.s.c. in the H^c-topology. These are easy exercises which use the geometric properties of the H^c-convergence and of the capacity.

Proposition 5.1.1 *The following classes of domains (defined in the beginning of chapter 4) are γ-compact:* $\mathcal{A}_{convex}$, $\mathcal{A}_{unif\,cone}$, $\mathcal{A}_{unif\,flat\,cone}$, $\mathcal{A}_{cap\,density}$, $\mathcal{A}_{unif\,Wiener}$; *in two dimensions the class* $\mathcal{A}_l$.

Exercise 1 Prove that if (Ω_n) is a sequence of convex sets converging in the H^c-topology to Ω, then Ω is also convex.

Exercise 2 Prove that the classes of domains satisfying a uniform exterior cone or flat cone condition are compact in the H^c-topology.

Hint Suppose that $\Omega_n \overset{H^c}{\to} \Omega$. For every $x \in \partial\Omega$ there exists a subsequence of (Ω_n) (still denoted with the same indices) and $x_n \in \partial\Omega_n$ such that $x_n \to x$. The condition in x for Ω is satisfied with the cone obtained as Hausdorff limit of the sequence of cones corresponding to the points x_n for Ω_n.

Exercise 3 Prove that the classes of domains satisfying a density capacity condition or a uniform Wiener criterion are compact in the H^c-topology.

Hint Prove that if $\Omega_n \overset{H^c}{\to} \Omega$ then

$$\text{cap}(\Omega^c \cap B_{x,t}, B_{x,2t}) \geq \limsup_{n \to \infty} \text{cap}(\Omega_n^c \cap B_{x,t}, B_{x,2t}),$$

for every $x \in \mathbb{R}^N$ and every $t > 0$ (see the necessary and sufficient conditions for the γ-convergence at the end of chapter 4).

Exercise 4 Prove that if (K_n) is a sequence of compact connected sets converging in the Hausdorff topology to K then K is also connected.

The direct methods of the calculus of variations and Proposition 5.1.1 give the following.

Theorem 5.1.2 *Let $J : D \times \mathbb{R}^{N+1} \to \mathbb{R}$ be a Carathéodory function. Then the shape optimization problem*

$$\min\{\int_\Omega j(x, u_{\Omega,f}, \nabla u_{\Omega,f}) : \Omega \in \mathcal{U}_{ad}\}$$

has at least one solution for $\mathcal{U}_{ad} = \mathcal{A}_{convex}$, $\mathcal{A}_{unif\,cone}$, $\mathcal{A}_{unif\,flat\,cone}$, $\mathcal{A}_{cap\,density}$, $\mathcal{A}_{unif\,Wiener}$, $\mathcal{A}_l$ *(for $N = 2$), respectively.*

5.2 A general abstract result for monotone costs

In this section we present a general framework in which the minimization problem of a monotone functional can be set.

Consider an ordered space $(X, \leq)$ and a functional $F : X \to \overline{\mathbb{R}}$. Suppose that X is endowed with two convergences denoted by γ and $w\gamma$, the last convergence being weaker than the first and sequentially compact (it will be called *weak gamma*). Moreover suppose that the functional F is γ lower semicontinuous. The relation we assume between γ and $w\gamma$ is the following one:

Assumption (A) *For every $x_n \overset{w\gamma}{\to} x$ there exist a sequence of integers $\{n_k\}$ and a sequence $\{y_{n_k}\}$ in X such that $y_{n_k} \leq x_{n_k}$ and $y_{n_k} \overset{\gamma}{\to} x$.*

The monotonicity of F becomes an important assumption because of the following result.

Proposition 5.2.1 *If $F : X \to \overline{\mathbb{R}}$ is monotone increasing and γ lower semicontinuous, then F is $w\gamma$ lower semicontinuous.*

Proof Let us consider $x_n \overset{w\gamma}{\to} x$, and let $\{x_{n_k}\}$ be a subsequence such that

$$\lim_{k \to \infty} F(x_{n_k}) = \liminf_{n \to \infty} F(x_n).$$

Using assumption (A) above there exists a subsequence (which we still denote by $\{x_{n_k}\}$) and $y_{n_k} \leq x_{n_k}$ such that $y_{n_k} \overset{\gamma}{\to} x$. The γ lower semicontinuity of F gives

$$F(x) \leq \liminf_{k \to \infty} F(y_{n_k})$$

and the monotonicity of F gives $F(y_{n_k}) \leq F(x_{n_k})$. Therefore

$$F(x) \leq \liminf_{k \to \infty} F(y_{n_k}) \leq \liminf_{k \to \infty} F(x_{n_k}) = \liminf_{n \to \infty} F(x_n)$$

which concludes the proof. ∎

Consider now another functional $\Phi : X \to \overline{\mathbb{R}}$ and the following problem

(5.1) $$\min\{F(x) : x \in X, \Phi(x) \leq 0\};$$

then we have

Theorem 5.2.2 *Let F be an increasing γ lower semicontinuous functional and let Φ be $w\gamma$ lower semicontinuous. Under the assumption (A) above, problem (5.1) admits at least one solution.*

Proof The proof follows straightforward by the directs method of the calculus of variations taking into account Proposition 5.2.1 and the fact that $w\gamma$ is supposed sequentially compact.   ∎

The general framework introduced above, even if quite trivial, applies very well in the case of shape optimization problems, and we shall apply it also in the case of obstacles. The main difficulty is to *"identify" the $w\gamma$-convergence, and to prove that assumption (A) is fulfilled.*

5.3   The weak γ-convergence for quasi-open domains

We use the general frame introduced in Section 5.2 for monotone functionals and introduce the *weak γ-convergence* for quasi-open sets.

Let us set $\mathcal{A} = \{A \subseteq D : A \text{ is quasi-open}\}$.

Definition 5.3.1 *We say that a sequence (A_n) in $\mathcal{A}$ weakly γ-converges to $A \in \mathcal{A}$ if w_{A_n} converges weakly in $H_0^1(D)$ to a function w such that $A = \{w > 0\}$ (here w is supposed to be quasi-continuous).*

We point out that, in general, the function w in Definition 5.3.1 does not coincide with w_A (this happens only if A_n γ-converges to A). Moreover, if if A_n weakly γ-converges to A then the Sobolev space $H_0^1(A)$ contains all the weak limits of sequences of elements of $H_0^1(A_n)$. Indeed, by [64, Lemma 3.2], if $u_n \in H_0^1(A_n)$ converge to u weakly in $H_0^1(\Omega)$, then $u = 0$ q.e. on $\{w = 0\}$, which gives $u \in H_0^1(A)$.

Lemma 5.3.2 *For every $A \in \mathcal{A}$ we have $\text{cap}\big\{A \triangle \{w_A > 0\}\big\} = 0$.*

Proof Since $w_A = 0$ q.e. on $\overline{D} \setminus A$, the inclusion $\{w_A > 0\} \subset A$ q.e. is obvious. In order to show the inclusion $A \subset \{w_A > 0\}$ q.e., by using [90, Lemma 1.5] we may find an increasing sequence (v_n) of nonnegative functions in $H_0^1(\Omega)$ such that $\sup v_n = 1_A$ q.e.; moreover, by [100, Proposition 5.5] for every v_n there exists a sequence $(\phi_{n,k})$ in $C_c^\infty(D)$ such that $\phi_{n,k} w_A$ tends to

v_n strongly in $H_0^1(D)$ and q.e. Therefore, since $\phi_{n,k}w_A = 0$ on $\{w_A = 0\}$, we also have $v_n = 0$ q.e. on $\{w_A = 0\}$ and so $1_A = 0$ q.e. on $\{w_A = 0\}$, which shows the inclusion $A \subset \{w_A > 0\}$ q.e.. ∎

Proposition 5.3.3 *If (A_n) γ-converges to A, then (A_n) also weakly γ-converges to A.*

Proof It follows from the definitions of γ-convergence and weak γ-convergence, by using Lemma 5.3.2 ∎

Proposition 5.3.4 *The weak γ-convergence on $\mathcal{A}$ is sequentially compact.*

Proof If (A_n) is a sequence in $\mathcal{A}$, by the boundedness of D we obtain that w_{A_n} is bounded in $H_0^1(D)$; hence we may extract a subsequence weakly converging in $H_0^1(D)$ to some function w. Defining $A = \{w > 0\}$ we get that a subsequence of A_n weakly γ-converges to A. ∎

Assumption (A) for the γ and the the $w\gamma$ convergences of quasi-open sets is contained in the following lemma.

Lemma 5.3.5 *Let $(X_n), X, Y$ in $\mathcal{A}$ be such that X_n weakly γ-converge to X and $X \subset Y$. Then there exist a subsequence $\{X_{n_k}\}$ of (X_n) and a sequence $\{Y_k\}$ in $\mathcal{A}$ such that $X_{n_k} \subset Y_k$ and Y_k γ-converge to Y.*

Proof This is a consequence of Lemma 4.8.3 for $p = 2$. ∎

Proposition 5.3.6 *The Lebesgue measure is weakly γ-lower semicontinuous on $\mathcal{A}$.*

Proof If $A_n \to A$ in the weak γ-sense, we have $w_{A_n} \to w$ weakly $H_0^1(\Omega)$, with $A = \{w > 0\}$, and for a subsequence the convergence is pointwise a.e. If $x \in A$ is such that $w_A(x) > 0$ and $w_{A_n}(x) \to w(x)$, then $w_{A_n}(x) > 0$ for n large enough, which implies that $x \in A_n$ for n large enough. Therefore, by Fatou's lemma,

$$|A| \leq \liminf_{n \to +\infty} |A_n|.$$

∎

5.4   Examples of monotone costs

Theorem 5.4.1 *Let $F : \mathcal{A} \to \overline{\mathbb{R}}$ be a function which is γ-lower semicontinuous and monotone decreasing with respect to inclusions. Then the optimization problem*

$$\min\{F(A) : |A| = m, A \in \mathcal{A}\}$$

admits at least a solution in $\mathcal{A}$.

Proof   This is a consequence of Theorem 5.2.2 and Lemma 5.3.5.   ∎

Example 5.4.2 (Domains with minimal k^{th} eigenvalue) For   every $A \in \mathcal{A}$ let $\lambda_k(A)$ be the kth eigenvalue of the Dirichlet Laplacian on $H_0^1(A)$, with the convention $\lambda_k(A) = +\infty$ if $\text{cap}(A) = 0$. It is well known that the mappings $A \mapsto \lambda_k(A)$ are decreasing with respect to set inclusion (see, e.g., Courant & Hilbert [87]). They are moreover continuous with respect to γ-convergence, so that Theorem 5.4.1 applies and for every $k \in \mathbb{N}$ and $0 \le c \le |\Omega|$ we obtain that the minimum

$$\min\{\lambda_k(A) : A \in \mathcal{A},\ |A| = c\}$$

is achieved. More generally, the minimum

$$\min\{\Phi(\lambda(A)) : A \in \mathcal{A},\ |A| = c\}$$

is achieved, where $\lambda(A)$ denotes the sequence $(\lambda_k(A))$ and the function $\Phi : \overline{\mathbb{R}}^{\mathbb{N}} \to \overline{\mathbb{R}}$ is lower semicontinuous and increasing, in the sense that

$$\lambda_k^h \to \lambda_k \quad \forall k \in \mathbb{N} \quad \Rightarrow \quad \Phi(\lambda) \le \liminf_{h \to \infty} \Phi(\lambda^h),$$

$$\lambda_k \le \mu_k \quad \forall k \in \mathbb{N} \quad \Rightarrow \quad \Phi(\lambda) \le \Phi(\mu).$$

Example 5.4.3 (Domains with minimal capacity) Since $\text{cap}(E)$ is increasing, the mapping $A \mapsto F(A) := \text{cap}(D \setminus A)$ is decreasing with respect to set inclusion. Since F is also Γ-continuous, as we may easily verify, Theorem 5.4.1 applies, so that the minimum

$$\min\{F(A) : A \in \mathcal{A},\ |A| = c\}$$

is achieved. If $\mathcal{F}(D)$ denotes the class of all quasi-closed subsets of Ω, we see immediately that the problem

(5.2) $$\min\big\{\mathrm{cap}(E) : E \in \mathcal{F}(D), \ |E| = k\big\}$$

admits at least a solution E_0 (it is enough to take $c = |\Omega| - k$ in the previous problem). Let us prove that

(5.3) $$\mathrm{cap}(E_0) = \min\big\{\mathrm{cap}(E) : E \subset \Omega, \ |E| = k\big\}.$$

For every subset E of Ω there exists a quasi-closed set E' such that $E \subset E'$ and $\mathrm{cap}(E) = \mathrm{cap}(E')$ (see, e.g., Fuglede [121, Section 2], or Dal Maso [92, Proposition 1.9]). If $|E| = k$, then $|E'| \geq k$, so that there exists $E'' \in \mathcal{F}(D)$ with $E'' \subset E'$ and $|E''| = k$. By 5.2 we have $\mathrm{cap}(E_0) \leq \mathrm{cap}(E'') \leq \mathrm{cap}(E') = \mathrm{cap}(E)$, which proves (5.3).

Example 5.4.4 (Domains which minimize an integral functional)
Let $f \in H^{-1}(\Omega)$, with $f \geq 0$, and let $g \colon \Omega \times \mathbb{R} \to \overline{\mathbb{R}}$ be a Borel function such that $g(x,\cdot)$ is lower semicontinuous and decreasing on $\mathbb{R}$ for a.e. $x \in \Omega$, and $g(x,s) \geq -\alpha(x) - \beta s^2$ for a suitable function $\alpha \in L^1(\Omega)$ and for a suitable constant $\beta \in \mathbb{R}$. For every $A \in \mathcal{A}$ let $u_A = R_A(f)$ and let

$$F(A) = \int_\Omega g\big(x, u_A(x)\big)\, dx.$$

Then F is lower semicontinuous with respect to γ-convergence and, since $f \geq 0$, the maximum principle and the monotonicity properties of g imply that F is decreasing with respect to set inclusion. Therefore, by Theorem 5.4.1 the minimum problem

$$\min\Big\{\int_\Omega g\big(x, u_A(x)\big)\, dx \ : \ A \in \mathcal{A}, \ |A| = c\Big\}$$

admits at least a solution.

5.5 The problem of optimal partitions

We consider shape cost functionals $F : \mathcal{A}^k \to [0, +\infty]$ and we study the following optimization problems:

$$\min\big\{F(A_1, \ldots, A_k) : \ A_i \in \mathcal{A}, \ A_i \cap A_j = \emptyset \ \text{for } i \neq j\big\}$$

where k is a fixed positive integer.

We say that F is γ-lower semicontinuous if for all γ-convergent sequences $A_i^n \to A_i$ for $i = 1, \ldots, k$ we have

$$(5.4) \qquad F(A_1, \ldots, A_k) \le \liminf_{n \to +\infty} F(A_1^n, \ldots, A_k^n).$$

Analogously, we say that F is weakly γ-lower semicontinuous if (5.4) holds for all sequences A_i^n which weakly γ-converge to A_i for $i = 1, \ldots, k$.

We say that F is monotonically decreasing (in the sense of the set inclusion) if for all $(A_1, \ldots, A_k)$, $(B_1, \ldots, B_k) \in \mathcal{A}^k$ such that $A_i \subset B_i$ for $i = 1, \ldots, k$ in the sense of capacity, i.e. $\mathrm{cap}(A_i \setminus B_i) = 0$, then

$$F(B_1, \ldots, B_k) \le F(A_1, \ldots, A_k).$$

We formulate now our existence result for shape optimization problems.

Theorem 5.5.1 *Let $F : \mathcal{A}^k \to [0, +\infty]$ be a weak γ-lower semicontinuous shape functional. Then the following optimization problem admits a solution:*

$$\min \Big\{ F(A_1, \ldots, A_k) \ : \ A_i \in \mathcal{A}, \ \mathrm{cap}(A_i \cap A_j) = 0 \Big\}.$$

Proof   Consider a minimizing sequence $(A_1^n, \ldots, A_k^n)_{n \in \mathbb{N}}$. Since the weak γ-convergence is sequentially compact, there exists a subsequence (still denoted with the same indices) such that

$$A_i^n \to A_i \qquad (i = 1, \ldots, k) \qquad \text{in the weak } \gamma\text{-sense.}$$

Since F is weakly γ-lower semicontinuous we have

$$F(A_1, \ldots, A_k) \le \liminf_{n \to +\infty} F(A_1^n, \ldots, A_k^n).$$

It remains only to prove that $(A_1, \ldots, A_k)$ satisfies the constraint, that is $\mathrm{cap}(A_i \cap A_j) = 0$ for $i \ne j$. We have that $w_{A_i^n} \cdot w_{A_j^n} = 0$ a.e. on Ω and $w_{A_i^n} \to w_i$ strongly in L^2. Therefore, $w_i \cdot w_j = 0$ a.e. on Ω. Since w_i and w_j are quasi-continuous functions, their product $w_i \cdot w_j$ is quasi-continuous too. Following [133] a quasi-continuous function which vanishes almost everywhere on an open set vanishes quasi-everywhere. So $w_i \cdot w_j = 0$ q.e. on Ω and so $\mathrm{cap}(A_i \cap A_j) = 0$. ∎

Corollary 5.5.2 *If $F : \mathcal{A}^k \to [0, +\infty]$ is monotonically decreasing and γ-lower semicontinuous, then problem (5.5.2) admits a solution.*

As an example, we may consider a cost functional J of the form

$$J(A_1, \ldots, A_k) = \phi\big(\lambda_{j_1}(A_1), \ldots, \lambda_{j_k}(A_k)\big)$$

where $\lambda_j(A)$ are the eigenvalues of the Laplace operator $-\Delta$ on $H_0^1(A)$, $j_1, \ldots, j_k$ are given positive integers, and $\phi(t_1, \ldots, t_k)$ is lower semicontinuous and nondecreasing in each variable. Then J fulfills all the assumptions of Corollary 5.5.2, so that the minimization problem

$$\min\Big\{ J(A_1, \ldots, A_k) \ : \ A_i \in \mathcal{A}, \ \text{cap}(A_i \cap A_j) = 0 \Big\}$$

has a solution. For instance this is the case of problem

$$\min\Big\{ \lambda_1(A_1) + \lambda_1(A_2) \ : \ A_i \in \mathcal{A}, \ \text{cap}(A_1 \cap A_2) = 0 \Big\}.$$

Using the $w\gamma$-l.s.c. of the Lebesgue measure, we obtain existence results for shape optimization problems like

$$\min\Big\{ J(A) + |A| \ : \ A \in \mathcal{A} \Big\}$$

with $J : \mathcal{A} \to [0, +\infty]$ weakly γ-semicontinuous, or more generally for

$$\min\Big\{ J(A, |A|) \ : \ A \in \mathcal{A} \Big\}$$

with $J : \mathcal{A} \times \mathbb{R} \to [0, +\infty]$ lower semicontinuous with respect to the $\{\text{weak } \gamma\} \times \{\text{euclidean}\}$-convergence and nondecreasing in the second variable.

5.6 Optimal obstacles

Again, in this section we discuss the nonlinear case. We consider a bounded open set D of $\mathbb{R}^N$ (for $N \geq 2$), a function $\psi \in W_0^{1,p}(D)$ with $p \in\,]1, +\infty[$, the family of admissible obstacles

$$X_\psi(D) = \{g : D \to \overline{\mathbb{R}} : g \leq \psi, \ g \text{ quasi upper semicontinuous}\},$$

and a cost functional $F : X_\psi(D) \to \overline{\mathbb{R}}$ which is monotone increasing, i.e. for all $g_1, g_2 \in X_\psi(D)$ with $g_1 \le g_2$ we have $F(g_1) \le F(g_2)$. Suppose that F is lower semicontinuous for the Γ-convergence of obstacle energy functionals (see Definition 5.6.2).

The result we are going to prove is the following (see [40]).

Theorem 5.6.1 *Under the assumptions above, for any constant $c \in \mathbb{R}$, the problem*

$$(5.5) \qquad \min\{F(g) : g \in X_\psi(D), \int_D g\,dx = c\}$$

admits a solution.

In order to use the general framework of monotone functionals we introduce the notions of γ and $w\gamma$-convergence for obstacles, we study their relationship and properties (compactness, assumption (A)) and prove Theorem 5.6.1.

The choice of obstacles as quasi-upper semicontinuous functions is natural, since one can replace an arbitrary obstacle by a suitable upper quasi semicontinuous one (see [17], [91]) such that the solution of problem (5.5) does not change.

For a quasi-upper semicontinuous function $g : D \to \overline{\mathbb{R}}$ we define the set

$$K_g = \{u \in W_0^{1,p}(D) : u \ge g \ \text{p-q.e.}\}$$

so that, for every $h \in L^q(D)$ (here q stands for the conjugate exponent of p, $1/p + 1/q = 1$), the solution of the obstacle problem associated to h and g is given by minimizing the associated energy

$$(5.6) \qquad \min\{\int_D \frac{1}{p}|\nabla u|^p dx - \int_D hu\,dx : u \in K_g\}.$$

For some fixed h the study of the solution of problem (5.6) when the obstacle g varies is done by the classical tool of the Γ-convergence related to the energy functional (see [17], [95]).

If (g_n) is a sequence of admissible obstacles and if $\{K_{g_n}\}$ converges in the sense of Mosco to K_g in $W_0^{1,p}(D)$ then it easy to see that the solutions u_n of problem (5.6) associated to g_n converge weakly in $W_0^{1,p}(D)$ to the solution of (5.6) corresponding to g. It is also well known that the Mosco convergence of the convex sets K_{g_n} to K_g is equivalent to the γ-convergence of the energy functionals associated to g_n and g.

Definition 5.6.2 *It is said that a sequence (g_n) of obstacles γ_o-converges to an obstacle g if the sequence of convex sets (K_{g_n}) converges to the convex set K_g in the sense of Mosco.*

In order to simplify the notations, in all this section when referring to obstacles we use the notation γ_o and when referring to quasi-open sets we add the index p, i.e. we use the notation γ_p. Of course, the γ_o-convergence of obstacles depends also on p, but since p is fixed, we omit it.

In order to use the abstract framework we have to introduce a second convergence $w\gamma_o$ on the class of admissible obstacles, and to prove that assumption (A) is fulfilled. The definition of the $w\gamma_o$-convergence for the obstacles will be given by means of the $w\gamma_p$-convergence of the level sets defined below (similar to the linear case).

In the sequel, w_A denotes as usual the solution of

$$\begin{cases} -\Delta_p w_A = 1 \text{ in } A \\ w_A \in W_0^{1,p}(A). \end{cases}$$

Definition 5.6.3 *We say that a sequence of quasi-open sets $\{A_n\}$ weakly γ_p-converges to a quasi-open set A if $w_{A_n} \rightharpoonup w$ weakly in $W_0^{1,p}(D)$ and $A = \{w > 0\}$. We say that a sequence of obstacles $(g_n)_{n \in \mathbb{N}}$ weak γ_o-converges to g (and we write $g_n \overset{w\gamma_o}{\to} g$) if there exists a dense set $D \subseteq \mathbb{R}$ such that*

$$\{g_n < t\} \overset{w\gamma_p}{\to} \{g < t\} \quad \forall t \in D.$$

Remark that this is a correct definition in the sense that if $g_n \overset{w\gamma_o}{\to} g$ and $g_n \overset{w\gamma_o}{\to} g'$ then $g = g'$. Thanks to this, there is no ambiguity in the choice of the dense set D.

For quasi-open sets, the γ_p-convergence is stronger than the $w\gamma_p$-convergence. While the γ_p-convergence is compact only in the family of relaxed domains, the $w\gamma_p$-convergence is compact in the family of quasi-open sets. In fact, if a sequence of quasi-open sets γ_p-converges to a measure μ, then this sequence $w\gamma_p$-converges to the regular set of this measure, i.e. the largest countable union of quasi-open sets with finite μ-measure (see [92, Definition 3.12]).

For obstacles, the relation between the γ_o-limit and the $w\gamma_o$-one, is not so simple to establish. Nevertheless, the γ_o-convergence of obstacles is stronger that the $w\gamma_o$-convergence (see Proposition 5.6.13), and the $w\gamma_o$-convergence is sequentially compact (see Lemma 5.6.4 below). Moreover, assumption (A) of the general framework is satisfied for the pair of topologies $(\gamma_o, w\gamma_o)$ and the classical order relation between functions in $X_\psi(D)$ (see Lemma 5.6.5 below). Indeed, we state the following lemmas.

Lemma 5.6.4 *For every sequence* $(g_n)_{n\in\mathbb{N}}$ *of elements of* $X_\psi(D)$ *there exist a subsequence* $\{g_{n_k}\}$ *and an obstacle* $g \in X_\psi(D)$ *such that* $g_{n_k} \xrightarrow{w\gamma_o} g$.

Lemma 5.6.5 *Consider a sequence of obstacles* $(g_n)_{n\in\mathbb{N}} \in X_\psi(D)$ *such that* $g_n \xrightarrow{w\gamma_o} g$. *There exist a subsequence* $(g_{n_k})_k$ *and a sequence* $(f_k)_k$ *with* $f_k \le g_{n_k}$ *such that* $f_{n_k} \xrightarrow{\gamma_o} g$.

In order to prove Theorem 5.6.1, we only remark the lower semicontinuity of the constraint.

Lemma 5.6.6 *Let* $g_n, g \in X_\psi(D)$, *such that* $g_n \xrightarrow{w\gamma_o} g$. *Then*

$$\int_D g\, dx \ge \limsup_{n\to\infty} \int_D g_n dx.$$

Proof [of theorem 5.6.1:] Consider a minimizing sequence $(g_n)_{n\in\mathbb{N}}$ of admissible obstacles. By Lemma 5.6.4 we may extract a subsequence (still denoted for simplicity by the same indices) which $w\gamma_o$-converges to some obstacle g in the sense of Definition 5.6.3. Since assumption (A) is fulfilled, by Proposition 5.2.1 we deduce

$$F(g) \le \liminf_{n\to\infty} F(g_n),$$

and by Lemma 5.6.6 on the upper semi-continuity of the constraint, we have $\int_D g\, dx \ge c$. If $\int_D g\, dx = c$ then the obstacle g is admissible and gives the minimum. If $\int_D g\, dx > c$ then the new obstacle $\tilde{g}$ defined by

$$\tilde{g}(x) = g(x) - \frac{1}{|D|}\left(\int_D g(y)dy - c\right)$$

is admissible (i.e. $\tilde{g} \in X_\psi(D)$ and $\int_D \tilde{g}\, dx = c$) and using the monotonicity of F we get $F(\tilde{g}) \le F(g)$. ∎

We point out the fact that in Theorem 5.6.1 the constraint is given by an equality, while in Theorem 5.2.2 the general form of the constraint is expressed by an inequality.

For the proof of lemmas 5.6.4, 5.6.5 and 5.6.6 we need some preparatory results. We recall first the following result of [91] which gives a characterization of the γ_o-convergence in terms of the behavior of the level sets and a technical result concerning the capacity of an increasing sequence of sets.

Lemma 5.6.7 *Let g_n, g be quasi-upper semicontinuous functions from D into $\overline{\mathbb{R}}$. Then $g_n \xrightarrow{\gamma_q} g$ and $K_g \neq \emptyset$ if and only if*

1. *there exists $T \subseteq]0, +\infty[$ with $0 \in \overline{T}$ such that for every $t \in T$*

$$\lim_{n \to \infty} \text{Cap}_p(\{g_n > t\}) = \text{Cap}_p(\{g > t\});$$

2. *there exist a dense set $D \subseteq \mathbb{R}$ and a family $\mathcal{B} \subseteq \mathcal{P}(D)$ such that for every $t \in D$ and every $B \in \mathcal{B}$*

$$\lim_{n \to \infty} \text{Cap}_p(\{g_n > t\} \cap B) = \text{Cap}_p(\{g > t\} \cap B);$$

the family $\mathcal{B}$ can be chosen dense in the sense of [92];

3. $\displaystyle \lim_{t \to \infty} \limsup_{n \to \infty} \int_t^{+\infty} \text{Cap}_p(\{g_n > s\})(s - t)^{p-1} ds = 0;$

4. $\displaystyle \lim_{t \to 0} \limsup_{n \to \infty} \int_0^t \text{Cap}_p(\{g_n > s\}) s^{p-1} ds = 0.$

Note that when replacing the metaharmonic capacity Cap_p with the harmonic one cap_p in D in the previous type of results, the analogous of Lemma 5.6.7 may fail to be true (see [91, Theorems 5.6, 5.9]).

Lemma 5.6.8 *For every $B \subseteq D$ and for every function g quasi-upper semicontinuous there exists an at most countable set $D(B)$ in $\mathbb{R}$ such that*

$$\text{cap}_p(\{g > t\} \cap B) = \text{cap}_p(\{g \geq t\} \cap B) \quad \forall t \in \mathbb{R} \setminus D(B).$$

Proof Fix $B \subseteq D$ and set for all $t \in \mathbb{R}$

$$U_t = \{g > t\} \cap B, \quad \tilde{U}_t = \{g \geq t\} \cap B.$$

We have $U_t \subseteq \tilde{U}_t$, from which it follows

(5.7) $$\text{cap}_p(U_t) \leq \text{cap}_p(\tilde{U}_t) \quad \forall t \in \mathbb{R}.$$

The function $t \mapsto \operatorname{cap}_p(\tilde{U}_t)$ is decreasing in t, so it is continuous on $N = \mathbb{R} \setminus D(B)$, with $D(B) \subseteq \mathbb{R}$ at most countable. Let us fix $\tau \in N$. For each $t \in \mathbb{R}$ such that $\tau < t$, we have $\operatorname{cap}_p(U_\tau) \geq \operatorname{cap}_p(U_t)$. Making $t \to \tau$ we have

$$(5.8) \qquad \operatorname{cap}_p(U_\tau) \geq \operatorname{cap}_p(\tilde{U}_\tau) \quad \forall \tau \in N.$$

Now from (5.7) and (5.8) we deduce $\operatorname{cap}_p(U_\tau) = \operatorname{cap}_p(\tilde{U}_\tau)$ which concludes the proof. ∎

Proofs of the main properties of γ_o and $w\gamma_o$-convergences. The following result is a very useful characterization of the γ_o-convergence of a sequence of obstacles in terms of the behavior of the level sets. Precisely it states that a sequence of obstacles is γ_o-convergent if and only if their level sets are γ_p-convergent.

Theorem 5.6.9 *Let $g_n, g \in X_\psi(D)$. Then $g_n \overset{\gamma_o}{\to} g$ if and only if there exists a family $T \subseteq \mathbb{R}$ such that $\mathbb{R} \setminus T$ is at most countable and*

$$\{g_n < t\} \overset{\gamma_p}{\to} \{g < t\} \quad \forall t \in T.$$

Proof The proof is based on Lemma 5.6.7 and relation (4.41). Let us suppose that $g_n \overset{\gamma_o}{\to} g$. Following Lemma 5.6.7 there exists a dense set $T \subseteq \overline{\mathbb{R}}$ with $\overline{\mathbb{R}} \setminus T$ at most countable, and a countable dense family $\mathcal{B} \subseteq \mathcal{P}(D)$ such that

$$\lim_{n \to \infty} \operatorname{cap}_p(\{g_n > t\} \cap B) = \operatorname{cap}_p(\{g > t\} \cap B)$$

for every $t \in T$ and every $B \in \mathcal{B}$. Using Lemma 5.6.8 for any $B_k \in \mathcal{B}$ and for any $n \in \mathbb{N}$ the family of $t \in \mathbb{R}$ such that

$$\operatorname{cap}_p(\{g_n > t\} \cap B_k) \neq \operatorname{cap}_p(\{g_n \geq t\} \cap B_k)$$

is at most countable. Therefore, eliminating all $t \in T$ for all $k, n \in \mathbb{N}$ such that the previous relation holds, one can find a set T' such that $\mathbb{R} \setminus T'$ is at most countable and such that

$$\lim_{n \to \infty} \operatorname{cap}_p(\{g_n \geq t\} \cap B) = \operatorname{cap}_p(\{g \geq t\} \cap B)$$

for every $t \in T'$ and every $B \in \mathcal{B}$. Using relation (4.41) we have that for all $t \in T'$ $\{g_n < t\} \overset{\gamma_p}{\to} \{g < t\}$. Since T' is dense in $\overline{\mathbb{R}}$ we conclude the necessity.

Suppose now that for a dense family $T \subseteq \mathbb{R}$ we have $\{g_n < t\} \stackrel{\gamma_p}{\to} \{g < t\}$. We prove that conditions 1), 2), 3), 4) of Lemma 5.6.7 are satisfied. From the fact that $g_n, g \leq \psi$ conditions 3) and 4) are satisfied. Following relation (4.41) condition 2) is also satisfied, by eliminating again an at most countable family of $t \in \mathbb{R}$ such that $\text{cap}_p(\{g_n > t\} \cap B_k) \neq \text{cap}_p(\{g_n \geq t\} \cap B_k)$. It remains to prove 1).

Let us fix some $t > 0$ and set $K_t = \{\psi \geq t\}$ which is a quasi-closed subset of D. Since $g_n \leq \psi$ we get

$$\text{cap}_p(\{g_n \geq t\} \cap K_t) = \text{cap}_p(\{g_n \geq t\}).$$

The idea is to find a set $B \in \mathcal{B}$ "between" K_t and D; this is not immediately possible since K_t is not closed but only quasi-closed, nevertheless, for any $\varepsilon > 0$ there exists a closed set $K_\varepsilon \subseteq K_t$ such that $\text{cap}_p(K_t \setminus K_\varepsilon) < \varepsilon$. Then

$$|\text{cap}_p(\{g_n \geq t\} \cap K_t) - \text{cap}_p(\{g_n \geq t\} \cap K_\varepsilon)| \leq \text{cap}_p(K_t \setminus K_\varepsilon) < \varepsilon.$$

Choosing a set $B \in \mathcal{B}$ such that $K_\varepsilon \subseteq B \subseteq D$ and for which

$$\text{cap}_p(\{g_n > t\} \cap B) \to \text{cap}_p(\{g > t\} \cap B)$$

we have

$$|\text{cap}_p(\{g > t\}) - \text{cap}_p(\{g_n > t\})| = |\text{cap}_p(\{g > t\} \cap K_t) - \text{cap}_p(\{g_n > t\} \cap K_t)|$$

$$\leq |\text{cap}_p(\{g > t\} \cap K_t) - \text{cap}_p(\{g > t\} \cap B)|$$

$$+ |\text{cap}_p(\{g > t\} \cap B) - \text{cap}_p(\{g_n > t\} \cap B)|$$

$$+ |\text{cap}_p(\{g_n > t\} \cap K_t) - \text{cap}_p(\{g_n > t\} \cap B)|.$$

The first and the last term of the right hand side are less than ε by the choice of B, and the middle term vanishes for $n \to \infty$. Hence we get

$$\lim_{n \to +\infty} \text{cap}_p(\{g_n > t\} = \text{cap}_p(\{g > t\}).$$

Therefore, point 1) of Lemma 5.6.7 is also proved and so the proof is concluded. ∎

Remark 5.6.10 The family T in the previous theorem can be replaced by a smaller one, as for example a dense set in $\mathbb{R}$.

Corollary 5.6.11 *Let* $g_n, g \in X_\psi(D)$. *Then* $g_n \overset{\gamma_\mathbb{R}}{\to} g$ *if and only if there exists a countable dense family* $\mathcal{D} \subseteq \mathbb{R}$ *such that*

$$(5.9) \qquad \{g_n < t\} \overset{\gamma_\mathbb{R}}{\to} \{g < t\} \quad \forall t \in \mathcal{D}.$$

Proof For the necessity see the first step of Theorem 5.6.9. Conversely, suppose that (5.9) holds for a set $\mathcal{D} \subseteq \mathbb{R}$ countable and dense. We prove that (5.9) holds for $t \in T$ where $\mathbb{R} \setminus T$ is at most countable (which implies, by Theorem 5.6.9, that $g_n \overset{\gamma_\mathbb{R}}{\to} g$).

For every $t \in \mathbb{R}$, possibly passing to subsequences, we have $w_{\{g_{n_k} < t\}} \rightharpoonup u_t$ weakly in $W_0^{1,p}(D)$ for a suitable function u_t. It will be enough to prove that $u_t = w_{\{g < t\}}$ up to an at most countable set. By assumption, $\mathcal{D}$ is dense in $\mathbb{R}$ and so we can find $t_1, t_2 \in \mathcal{D}$ with $t_1 \leq t \leq t_2$. Then

$$(5.10) \qquad w_{\{g < t_1\}} \leq u_t \leq w_{\{g < t_2\}}.$$

Let us define

$$A_\tau = \{g < \tau\} \text{ for any } \tau \in \mathcal{D}, \quad \tau \leq t,$$

$$B_\tau = \{g < \tau\} \text{ for any } \tau \in \mathcal{D}, \quad \tau \geq t.$$

The sets A_τ and B_τ are quasi-open and we have that $\{A_\tau\}_{\tau \in \mathcal{D}, \tau \leq t}$ is increasing and $\{B_\tau\}_{\tau \in \mathcal{D}, \tau \geq t}$ is decreasing with respect to the set inclusions. Now, from the theory of the γ_p-convergence (see [92], [140]), we have

$$(5.11) \qquad A_\tau \overset{\gamma_\mathbb{R}}{\to} \bigcup_{\tau \in \mathcal{D}, \tau \leq t} A_\tau = A_t \quad \text{as } \tau \to t$$

and

$$(5.12) \qquad w_{\{g < t\}} = \sup\{w_{\{g < \tau\}} : \tau \in \mathcal{D}, \tau \leq t\}.$$

One has to pay attention to the fact that an arbitrary union of quasi-open sets is not generally quasi-open. In this case, the monotonicity plays an essential role. In fact any increasing sequence of quasi-open sets is γ_p-convergent to their union. Moreover

$$(5.13) \qquad \{g < t\} \subseteq B_\tau \quad \forall \tau \in \mathcal{D}, \tau \geq t,$$

then

(5.14) $$w_{\{g<t\}} \le w_{\{g<\tau\}} \quad \forall \tau \in \mathcal{D}, \ \tau \ge t,$$

from which it follows

(5.15) $$w_{\{g<t\}} \le \inf\left\{w_{\{g<\tau\}} : \tau \in \mathcal{D}, \ \tau \ge t\right\}.$$

Now from (5.10), ... ,(5.15) we have

(5.16) $$w_{\{g<t\}} = \sup\left\{w_{\{g<\tau\}} : \tau \in \mathcal{D}, \ \tau \le t\right\}$$

$$\le u_t \le \inf\left\{w_{\{g<\tau\}} : \tau \in \mathcal{D}, \ \tau \ge t\right\}.$$

Let us consider the mapping $t \mapsto \|w_{\{g<t\}}\|_{L^p(D)}$ which is increasing, and hence it has an at most countable points of discontinuity. If t is a point where this mapping is continuous we have

$$\sup\left\{w_{\{g<\tau\}} : \tau \in \mathcal{D}, \ \tau \le t\right\} = \inf\left\{w_{\{g<\tau\}} : \tau \in \mathcal{D}, \ \tau \ge t\right\}$$

and so from (5.16) we have $u_t = w_{\{g<t\}}$, which concludes the proof. ∎

We list now the main properties of γ_o and $w\gamma_o$-convergences for the obstacles.

Proposition 5.6.12 *Let $g_n, g \in X_\psi(D)$. The following conditions are equivalent:*

(i) *there exists a dense family $\mathcal{D} \in \mathbb{R}$ such that $\{g_n < t\} \xrightarrow{w\gamma_p} \{g < t\}$ for all $t \in \mathcal{D}$;*

(ii) *there exists an at most countable family $N \in \mathbb{R}$ such that $\{g_n < t\} \xrightarrow{w\gamma_p} \{g < t\}$ for all $t \in \mathbb{R} \setminus N$.*

Proof If (ii) is true, obviously (i) follows. Conversely, let us suppose that (i) is true. For all $s \in \mathcal{D}$ we have $w_{\{g_n<t\}} \rightharpoonup w_s$ in $W_0^{1,p}(D)$, $\{w_s > 0\} = \{g < s\}$ and, from the upper semicontinuity of g, for all $t \in \mathbb{R}$,

$$\{g < t\} = \bigcup_{s \in \mathcal{D}, s \le t} \{g < s\}.$$

We define for any $t \in \mathbb{R}$ $w_t := \sup\limits_{s \in \mathcal{D}, s \le t} w_s$. From the monotonicity of the mapping $s \mapsto w_s$ the definition above is consistent and we have

$$\{w_t > 0\} = \bigcup_{s \in \mathcal{D}, s \le t} \{w_s > 0\} = \bigcup_{s \in \mathcal{D}, s \le t} \{g < s\} = \{g < t\}.$$

From the boundedness of $w_{\{g_n < t\}}$, possibly passing to a subsequence still denoted by the same indices, we have

$$w_{\{g_n < t\}} \rightharpoonup u_t.$$

It is sufficient to prove that there exists an at most countable set N such that $u_t(x) = w_t(x)$ p-q.e. for $t \in \mathbb{R} \setminus N$. We observe that, by monotonicity, for every $s \in \mathcal{D}$ with $s \le t$ we have

(5.17)
$$w_{\{g_n < s\}} \le w_{\{g_n < t\}},$$

so passing to the limit in (5.17) as $n \to +\infty$ we obtain

$$w_s \le u_t \quad \forall s \in \mathcal{D}, s \le t,$$

from which

(5.18)
$$u_t \ge \sup_{s \in \mathcal{D}, s \le t} w_s = w_t.$$

On the other hand, again by the monotonicity, we also have

$$u_t \le w_s \quad \forall s \in \mathcal{D}, \ s \ge t,$$

from which

(5.19)
$$u_t \le \inf_{s \in \mathcal{D}, s \ge t} w_s.$$

Now, denoting by $\tilde{w}_t$ the right hand side of (5.19), for $t \in \mathbb{R} \setminus N$, with N at most countable, we have $w_t = \tilde{w}_t$ since the mapping $t \mapsto \|w_t\|_{L^p(D)}$ is monotone increasing and it has only at most countable points of discontinuity. So we obtain from (5.18) and (5.19) that $u_t = w_t$ for every $t \in \mathbb{R} \setminus N$ which ends the proof. ∎

Proposition 5.6.13 *Suppose that $g_n \xrightarrow{\gamma_\bullet} g$. Then $g_n \xrightarrow{w\gamma_\bullet} g$.*

Proof It is an immediate consequence of Theorem 5.6.9 and Definition 5.6.3. ∎

We are now able to give the proof of Lemma 5.6.4 which states the compactness of the $w\gamma_o$-topology.

Proof [of lemma 5.6.4] Consider an enumeration $\{r_1, r_2, ...\}$ of the set $\mathbb{Q}$ of rational numbers. For the level r_1 there exists a subsequence of $(g_n)_{n \in \mathbb{N}}$ (still denoted with the same indices) such that $w_{\{g_n < r_1\}} \rightharpoonup w_{r_1}$ weakly in $W_0^{1,p}(D)$. For the level r_2 there exists a subsequence of the previous one such that $w_{\{g_n < r_2\}} \rightharpoonup w_{r_2}$ weakly in $W_0^{1,p}(D)$. In such a way for any $r_k \in \mathbb{Q}$ one can extract a subsequence (of the sequence established for r_{k-1}) such that $w_{\{g_n < r_k\}} \rightharpoonup w_{r_k}$ weakly in $W_0^{1,p}(D)$. By a diagonal procedure we can choose an element of the first sequence such that

$$\|w_{\{g_{n_1} < r_1\}} - w_{r_1}\|_{L^p(D)} < 1,$$

then a second element of the second sequence such that

$$\|w_{\{g_{n_2} < r_1\}} - w_{r_1}\|_{L^p(D)} < \frac{1}{2} \quad \text{and} \quad \|w_{\{g_{n_2} < r_2\}} - w_{r_2}\|_{L^p(D)} < \frac{1}{2},$$

and continuing this procedure, one construct a subsequence $(g_{n_i})_i$ such that

$$\forall k \in \mathbb{N} \quad w_{\{g_{n_i} < r_k\}} \overset{L^p(D)}{\to} w_{r_k} \quad \text{for} \quad i \to \infty.$$

This means exactly that for all $t \in \mathbb{Q}$

$$\{g_{n_i} < t\} \overset{w\gamma_p}{\to} \{w_t > 0\} \quad \text{for} \quad i \to \infty.$$

One define then the limit obstacle through its level sets

(5.20) $$\{g < t\} = \{w_t > 0\} \quad \text{for all} \quad t \in \mathbb{Q}.$$

If $t_1, t_2 \in \mathbb{Q}$ with $t_1 < t_2$ then obviously $\{g < t_1\} \subseteq \{g < t_2\}$. The function g is defined by

$$g(x) = \inf\{t \in \mathbb{Q} : x \in \{g < t\}\}.$$

Hence for every $t \in \mathbb{R} \setminus \mathbb{Q}$ we have

(5.21) $$\{g < t\} = \bigcup_{s \in \mathbb{Q}, s < t} \{g < s\}.$$

One can see that

(i) g is correctly defined.

(ii) g is quasi-upper semicontinuous, because its level sets $\{g < t\}$ are quasi-open, being by definition, countable unions of quasi-open sets (see (5.20) and (5.21)).

(iii) $g \in X_\psi(D)$, because $\{\psi < t\} \supseteq \{g < t\}$ for every $t \in \mathbb{R}$. Indeed, let us consider first the case $t \in \mathbb{Q}$. From hypothesis we have $g_{n_i} \leq \psi$, for all i, hence

$$\{\psi < t\} \supseteq \{g_{n_i} < t\}, \quad \forall i \in \mathbb{N},$$

from which

$$w_{\{g_{n_i} < t\}} \leq w_{\{\psi < t\}}.$$

Passing into the limit as $i \to +\infty$, we get

$$w_t \leq w_{\{\psi < t\}},$$

so that

$$\{w_{\{\psi < t\}} > 0\} = \{\psi < t\} \supseteq \{w_t > 0\} = \{g < t\}.$$

Fix now $t \in \mathbb{R} \setminus \mathbb{Q}$. We have

$$\{g < t\} = \bigcup_{s < t, s \in \mathbb{Q}} \{g < s\} \subseteq \bigcup_{s < t, s \in \mathbb{Q}} \{\psi < s\} \subseteq \{\psi < t\}$$

which shows that $g \in X_\psi(D)$.

It remains to prove that $g_{n_i} \overset{w\gamma_o}{\to} g$, or equivalently, that there exists a dense set $\mathcal{D} \subseteq \mathbb{R}$ such that

$$w_{\{g_{n_i} < t\}} \rightharpoonup w_t, \quad \{w_t > 0\} = \{g < t\}, \quad \forall t \in \mathcal{D}.$$

If $t \in \mathbb{Q}$ the above is true from the definition of g, which concludes the proof. ∎

In order to prove assumption (A) for the $w\gamma_o$ and γ_o-convergences, we give a technical result for domains, necessary for the construction of the γ_o-convergent sequence deriving from the $w\gamma_o$-convergent one.

We give now the proof of Lemma 5.6.5, which asserts that assumption (A) is satisfied for the couple of topologies $(\gamma_o, w\gamma_o)$ in $X_\psi(D)$.

Proof [of Lemma 5.6.5] There are two steps. We first consider the particular case of obstacles whose ranges are in a finite set. Let us consider the numbers

$$-\infty = l_1 < l_2 < \cdots < l_q$$

and the special family of obstacles

$$\mathcal{A}[l_1, \ldots, l_q] = \{g \in X_\psi(D) \ : \ g(x) \in \{l_1, \ldots, l_q\}\}.$$

We construct the sets

$$A_1(g) = \{g < l_2\}, \quad A_2(g) = \{g < l_3\}, \ldots, A_{q-1}(g) = \{g < l_q\}$$

which are quasi-open and such that $A_1 \subseteq A_2 \subseteq \ldots \subseteq A_{q-1}$. Consider now a sequence $(g_n)_{n \in \mathbb{N}} \in \mathcal{A}[l_1, \ldots, l_q]$ which weak γ_o-converges to a function g. To every function g_n we associate as before the sets $g_n \to A_1(g_n), \ldots, A_{q-1}(g_n)$ and, using the compactness of the weak γ_p-convergence for sets, we can write (for a subsequence still denoted with the same indices)

$$A_i(g_n) \overset{w\gamma_p}{\longrightarrow} A_i \quad \forall i = 1, \ldots, q-1.$$

By the definition of the weak γ_o-convergence we have

$$g = l_1 \text{ on } A_1, \quad g = l_2 \text{ on } A_2 \setminus A_1, \ldots, g = l_q \text{ on } D \setminus (A_1 \cup \cdots \cup A_{q-1}).$$

Moreover, g is quasi-upper semicontinuous and $g \le \psi$. For this last inequality it is sufficient to prove for any $i = 2, \ldots, q$ that $\{g < l_i\} \subseteq \{\psi < l_i\}$. This follows from the fact that $\{g_n < l_i\} \subseteq \{\psi < l_i\}$ and from the properties of the weak γ_p-convergence of sets.

We construct now the sequence f_n of admissible obstacles such that $f_n \overset{\gamma_o}{\to} g$ with $f_n \le g_n$. Using Lemma 4.8.3 for subsequences still denoted with the same indices there exist sets $G_i^n \supseteq A_i(g_n)$ such that $G_i^n \overset{\gamma_p}{\to} A_i$. For fixed n, the sets $G_1^n, \ldots, G_{q-1}^n$ are not ordered. Therefore one apply the nonlinear version of Proposition 4.5.4 and consider

(5.22)	$\tilde{G}_1^n = G_1^n \cap G_2^n \cap \cdots \cap G_{q-1}^n$
(5.23)	$\tilde{G}_2^n = G_2^n \cap G_3^n \cap \cdots \cap G_{q-1}^n$
(5.24)	$\ldots$
(5.25)	$\tilde{G}_{q-1}^n = G_{q-1}^n.$

Then $\tilde{G}_1^n \subseteq \tilde{G}_2^n \subseteq \ldots \subseteq \tilde{G}_{q-1}^n$ and $\tilde{G}_i^n \overset{\gamma_R}{\to} A_i$. Moreover, $\tilde{G}_i^n \supseteq A_i(g_n)$. Defining the obstacle f_n with $\tilde{G}_1^n, \ldots, \tilde{G}_{q-1}^n$ in the following way: $f_n = l_1$ on $\tilde{G}_1^n$, $f_n = l_2$ on $\tilde{G}_2^n \setminus \tilde{G}_1^n$, $\ldots$, $f_n = l_q$ on $D \setminus \tilde{G}_{q-1}^n$, we get that f_n is quasi-upper semicontinuous, $f_n \leq g_n$ and $f_n \overset{\gamma}{\to} g$.

In a second step of the proof, a general obstacle is approached by obstacles with a finite range. We define for any $k \in \mathbb{N}$ the family of levels

$$\mathcal{R}_k = \{l_1, \ldots, l_k\} \cup -\infty$$

where $(l_k)_k$ is a dense set in $\overline{\mathbb{R}}$ with the property that the weak γ_p-convergence holds on levels l_i. We have $\mathcal{R}_{k_1} \subseteq \mathcal{R}_{k_2}$ if $k_1 \leq k_2$. Denote by $\mathcal{R} = \bigcup_{k \in \mathbb{N}} \mathcal{R}_k$.

For some obstacle $g \in X_\psi(D)$ we define the truncation on $\mathcal{R}_k$ of g in the following way

$$T_k(g)(x) = \sup\{l \in \mathcal{R}_k : g(x) \geq l\}.$$

Obviously $T_k(g) \in \mathcal{A}[\mathcal{R}_k]$ and $T_k(g) \leq g$. Moreover, if $l \in \mathcal{R}_k$ then for all $k' \geq k$ we have

$$\{T_{k'}(g) < l\} = \{g < l\}.$$

As in the first step, for $k = 1$ there exists a subsequence, still denoted with the same indices, such that

$$T_1(g_n) \overset{w\gamma_o}{\to} T_1(g);$$

we consider a subsequence and a sequence $f_n^1 \leq T_1(g_n)$ with $f_n^1 \overset{\gamma_q}{\to} T_1(g)$. For $k = 2$ there exists (as in step one) a subsequence such that

$$T_2(g_n) \overset{w\gamma_o}{\to} T_2(g);$$

again, we consider a subsequence and a sequence $f_n^2 \leq T_2(g_n)$ with $f_n^2 \overset{\gamma_q}{\to} T_2(g)$. We continue this procedure for any $k \in \mathbb{N}$ and we chose a diagonal sequence $(f_{n_k}^k)_k$ with the property that $d_\gamma(f_{n_k}^k, T_k(g)) \leq 1/k$. Here by d_γ one denotes the distance which generates the same topology of the γ_o-convergence (see [89] for the metrizability of the γ_o-convergence). On the other hand, using Theorem 5.6.9 we have $T_k(g) \overset{\gamma_q}{\to} g$ and therefore we found a subsequence $(f_{n_k}^k)_k$ which satisfies the desired properties.   ∎

Since assumption (A) is fulfilled, the general framework presented in Section 2 and in particular Theorem 5.2.2 could be applied. Nevertheless, in the case of obstacles the integral constraint plays a very important role.

Lower semicontinuity of the constraint. In order to prove that an integral constraint of the type

$$\int_D g\,dx \geq c$$

is weak γ_o-compact, we give the following.

Proposition 5.6.14 *Let* $g_n, g \in X_\psi(D)$, *with* $g_n \xrightarrow{w\gamma_o} g$. *Then*

$$g(x) \geq \limsup_{n\to\infty} g_n(x) \quad \text{for a.e. } x \in D.$$

Proof Indeed, consider some $x \in D$ and $l \in \mathbb{R}$ with $g(x) < l$. From the weak γ_o-convergence, there exists some l' between $g(x)$ and l, such that

$$\{g_n < l'\} \xrightarrow{w\gamma_p} \{g < l'\};$$

then $1_{\{g<l'\}} \leq \liminf_{n\to\infty} 1_{\{g_n<l'\}}$, where we denote by 1_C the function such that $1_C(x) = 1$ if $x \in C$ and and $1_C(x) = 0$ otherwise. In particular, this means that $g_n(x) < l'$ for n large enough, that is

$$\limsup_{n\to\infty} g_n(x) \leq l'.$$

Since l was arbitrary, and $g(x) < l' < l$ the proof is concluded. ∎

We are now in the position to give the proof of Lemma 5.6.6.

Proof [of lemma 5.6.6] From Proposition 5.6.14 we have $g \geq \limsup_{n\to\infty} g_n$. The conclusion now follows by Fatou's lemma. ∎

Remark 5.6.15 A constraint of the type $\int_D \Phi(x, g(x))dx \geq c$ can replace the previous one in the general framework of Theorem 5.2.2 provided the function $\Phi(x, \cdot)$ is increasing and upper semicontinuous.

Remark 5.6.16 Given $h \in L^2(D)$, let us denote by u_g the solution of

$$\min\{\frac{1}{2}\int_D |\nabla u|^2 dx - \int_D hu\, dx \; : \; u \geq g \text{ in } D\}.$$

An interesting problem is to minimize the integral of u_g among all admissible obstacles, that is to solve

(5.26) $$\min\{\int_D u_g\, dx \; : \; g \in X_\psi(D), \int_D g\, dx = c\}.$$

We observe that the functional $g \to F(g) = \int_D u_g dx$ is increasing (from the general properties of the solution u_g) and γ_o lower semicontinuous, hence Theorem 5.6.1 applies.

We can also consider more general problems of the form

(5.27) $$\min\{\int_D \big[\int_\mathbb{R} f(x,t,u_g(x), \Lambda_I(\{g < t\}), \Lambda_J(\{u_g < t\})) \, dt\big]\, dx \; :$$

$$g \in X_\psi(D), \int_D g\, dx = c\},$$

where, for every finite set $I \subseteq \mathbb{N}$, $\Lambda_I(A) = (\lambda_i(A))_{i \in I}$, being $\lambda_i(A)$ the i-th eigenvalue of the Laplacian on A with Dirichlet boundary conditions, counted with its multiplicity. Here $f : D \times \mathbb{R} \times \mathbb{R} \times \mathbb{R}^I \times \mathbb{R}^J \to [0, +\infty]$ is a Borel function with $f(x,t,\cdot,\cdot,\cdot)$ increasing and lower semicontinuous. By the monotonicity and γ_2-continuity of mappings $A \to \lambda_i(A)$ (see [64] for details), Theorem 5.6.1 applies and we obtain the existence of at least one solution for problem (5.27).

A very special case is the minimization of the energy of the obstacle. Given $h \in L^q(D)$, the energy associated to the obstacle $g \in X(D)$ is given by

(5.28) $$\min_{u \in K_g} \tilde{F}(u) \quad \text{where} \quad \tilde{F}(u) = \frac{1}{p}\int_D |\nabla u|^p dx - \int_D hu\, dx.$$

In a more general framework, one can minimize a variational functional F of the type

$$F(g) = \min_{u \in K_g} \tilde{F}(u),$$

where $\tilde{F} : W_0^{1,p}(D) \to \overline{\mathbb{R}}$ is given. Of course, F has the monotonicity property, and moreover, if $\tilde{F}$ is weakly lower semicontinuous and coercive on $W_0^{1,p}(D)$, then F is γ_o-lower semicontinuous. Hence one can apply Theorem 5.6.1 in order to get the existence of a minimizer.

For this class of functionals F, one can give a direct proof of the existence of a minimizer, if the constraint on the admissible obstacle $g \leq \psi$ is dropped. In this case, the minimization problem

$$(5.29) \qquad \min \left\{ \tilde{F}(u) : u \in K_g, \, g \text{ quasi-u.s.c.}, \int_D g \, dx \geq c \right\}$$

is equivalent to the following one

$$(5.30) \qquad \min \left\{ \tilde{F}(u) : u \in W_0^{1,p}(D), \int_D u \, dx \geq c \right\}.$$

Indeed, if (g, u) is a solution for (5.29), then u is a minimizer for (5.30). Conversely if u^* minimizes for (5.30), then the couple (u^*, u^*) is a minimizer for (5.29). Hence the existence of a solution for problem (5.29) derives from a solution of problem (5.30), which in this case can also be obtained by the direct methods of the calculus of variations.

Using the equivalence between (5.29) and (5.30), one can write for problem (5.28) the necessary optimality conditions for the solution u^*. Performing the directional derivatives and using the Lagrange multiplier, we get for some positive constant λ

$$(5.31) \qquad -\Delta_p u^* = h + \lambda \quad \text{in } W_0^{1,p}(D).$$

The regularity of u^* depends then on the regularity of h, hence for quite smooth data h and D, the optimal solution of (5.28) belongs to the class of regular functions.

However, the existence method presented above does not apply to more general situations when the optimization problem is not written in the form (5.29), as for example problem (5.27).

Bilateral obstacles, non-linear shape optimization problems. By analogy, the previous results can be extended to the bilateral obstacle problem. In order to consider bilateral problems, we define the family

$$\tilde{X}_\psi(D) = \{g : D \to \overline{\mathbb{R}} : g \text{ quasi-l.s.c. } g \geq \psi\}$$

with ψ fixed in $W_0^{1,p}(D)$; the associated convex sets are

$$\tilde{K}_g = \{u \in W_0^{1,p}(D) : u \le g \ p\text{-q.e. }\}.$$

In this case, the γ_o-convergence, respectively $w\gamma_o$-convergence, are defined as in Definitions 5.6.2 and 5.6.3 by using the upper level sets: $\{g > t\}$. We denote them by $\tilde{\gamma}_o$, respectively $w\tilde{\gamma}_o$.

Then Theorem 5.6.1 still holds under the assumptions:

1. F is lower semicontinuous with respect to the $\tilde{\gamma}_o$-convergence;

2. F is monotone decreasing with respect to the usual order of functions.

We can now consider the case of bilateral problems. Fix first a function $\psi \in W_0^{1,p}(D)$; the admissible set is now

$$Y_\psi(D) = \{(g_1, g_2) : g_i : D \to \overline{\mathbb{R}}, i = 1, 2,$$

$$g_1 \text{ quasi-u.s.c., } g_2 \text{ quasi-l.s.c., } g_1 \le \psi \le g_2\}.$$

We define

$$K_{g_1,g_2} = \{u \in W_0^{1,p}(D) : g_1 \le u \le g_2 \ p\text{-q.e.}\}$$

and consider a functional $F : Y_\psi(D) \to \overline{\mathbb{R}}$. Then Theorem 5.6.1 still holds under the assumptions:

1. $F(.,.)$ is lower semicontinuous with respect to the $(\gamma_o, \tilde{\gamma}_o)$-convergence;

2. $F(.,.)$ is monotonously increasing with respect to the first variable and decreasing with respect to the second one.

In this case, the constraint is of the form

$$\int_D g_1 dx = c, \quad \int_D g_2 dx = \tilde{c},$$

with, of course, $c \le \tilde{c}$.

The shape optimization problem solved in Section 5.3 for $p = 2$ is a particular case of the previous result and follows directly considering the family of obstacles

$$\mathcal{O}(D) = \{(g_1, g_2) : g_1 = -\infty \cdot 1_A, \ g_2 = +\infty \cdot 1_A, \ A \subseteq D, \ A \text{ quasi-open}\}.$$

It is sufficient to see that $\mathcal{O}(D)$ is closed with respect to the $(w\gamma_o, w\tilde{\gamma}_o)$-convergence. In fact, a bilateral obstacle $(-\infty \cdot 1_A, +\infty \cdot 1_A)$ is identified with the quasi-open set A. In this case, the bilateral obstacle problem becomes a Dirichlet problem with homogeneous boundary condition associated to the p-Laplacian on the quasi-open set A.

Chapter 6

Optimization problems for functions of eigenvalues

6.1 Setting the problem

Minimization problems for eigenvalues are not new in the literature; since the first result by Krahn [145], [146] and Faber [115] concerning the minimality of a disk in $\mathbb{R}^2$ for the first eigenvalue of the Laplace operator $-\Delta$ with Dirichlet boundary conditions, among domains with equal area, many other results have been obtained.

Denoting by B_1 the ball of mass c, and by B_2 the union of two disjoint balls of mass $c/2$ for every bounded open set A of measure c the following inequalities hold:

- $\lambda_1(A) \geq \lambda_1(B_1)$ (proved by Faber, [115] and Krahn, [145], [146]);

- $\lambda_2(A) \geq \lambda_2(B_2) = \lambda_1(B_2)$ (we refer to Krahn, [146]; see also [168] for a proof by P. Szegö);

- $\lambda_2(B_1)/\lambda_1(B_1) \geq \lambda_2(A)/\lambda_1(A) \geq 1$ (proved by Ashbaugh and Benguria, see [13]);

Notice that by density arguments and γ-continuity properties, in the previous inequalities the set A can be replaced by an arbitrary (not necessarily bounded) quasi-open set of measure less than or equal to c.

We are concerned with problems of the form

$$(6.1) \qquad \min\{\Phi(\lambda(\Omega)) \ : \ |\Omega| \leq c, \Omega \in \mathcal{A}\}$$

where $\lambda(\Omega)$ denotes the sequence $(\lambda_j(\Omega))_j$ of all eigenvalues of the Laplace operator with Dirichlet boundary conditions, Φ is a given function, $m > 0$ is given and $\mathcal{A}$ is the class of admissible domains. A choice we can do is to take $\mathcal{A} = \{\Omega \subseteq D\}$ where the domain D (the so called "*design region*") is either a given, bounded open subset of $\mathbb{R}^N$, or $D = \mathbb{R}^N$. We shall see that the two cases are quite different, and most of the results obtained for the first cannot be easily extended to the second.

Following Theorem 5.4.1, a general existence result can be adapted for functionals depending on eigenvalues. We want to remark that the first assumption of Theorem 5.4.1 is verified for a large number of situations; for instance we have (see Buttazzo and Dal Maso [64])

$$\Omega_n \xrightarrow{\gamma} \Omega \qquad \Longrightarrow \qquad \lambda_j(\Omega_n) \to \lambda_j(\Omega) \quad \text{for all } j \in \mathbb{N}.$$

Therefore the cost functions

$$F(\Omega) = \Phi(\lambda(\Omega))$$

are γ-continuous as soon as the function $\Phi : \mathbb{R}^N \to [0, +\infty]$ is lower semicontinuous, in the sense that

$$(6.2) \qquad z_j^n \to z_j \text{ for all } j \in \mathbb{N} \implies \Phi(z) \le \liminf_{n \to \infty} \Phi(z^n).$$

The monotonicity assumption is, on the contrary, much more restrictive. However, in the case of problems involving eigenvalues, of the form (6.1), Theorem 5.4.1 includes the case of functions Φ which are monotone increasing, that is

$$(6.3) \qquad z_j^1 \le z_j^2 \text{ for all } j \in \mathbb{N} \implies \Phi(z^1) \le \Phi(z^2).$$

This relies on the well known fact that all the eigenvalues of an elliptic operator with Dirichlet boundary conditions are decreasing functions of the domain. Therefore, summarizing, from Theorem 5.4.1 we can deduce the following result.

Corollary 6.1.1 *Let $\Phi : \mathbb{R}^N \to [0, +\infty]$ be a function which is lower semicontinuous in the sense of (6.2), and monotone decreasing in the sense of (6.3). Then the optimization problem*

$$\min\{\Phi(\lambda(\Omega)) : |\Omega| = c, \Omega \in \mathcal{A}\}$$

admits at least a solution in $\mathcal{A}$.

In this chapter, we discuss two directions in which we weaken the hypotheses of the previous corollary. We consider *non-monotone functionals* Φ and *unbounded design regions*.

6.2 A short survey on continuous Steiner symmetrization

An important tool often used in shape optimization is the continuous Steiner symmetrization denoted CSS (see Brock [32]).

For $N \geq 1$ denote by $M(\mathbb{R}^N)$ the class of Lebesgue measurable subsets of $\mathbb{R}^N$.

Let us begin by the one dimensional case. A family of mappings

$$E_t : M(\mathbb{R}) \to M(\mathbb{R}), \ 0 \leq t \leq \infty,$$

is called a continuous symmetrization if it satisfies

1. $m(E_t(M)) = m(M)$.

2. If $M \subseteq N$ then $E_t(M) \subseteq E_t(N)$.

3. $E_t(E_s(M)) = E_{s+t}(M)$.

4. If $I = [a, b]$ is a closed interval, then $E_t(I) = [a^t, b^t]$ where

$$a^t = \frac{1}{2}(a - b + e^{-t}(a + b))$$
$$b^t = \frac{1}{2}(b - a + e^{-t}(a + b)).$$

In [32] the existence of such a family is proved. In the sequel, we will denote by $M^t = E_t(M)$, the continuous symmetrization of a set M. If M is a finite union of intervals, $M = (a_1, b_1) \cup (a_2, b_2) \cup ... \cup (a_k, b_k)$ such that $a_1 < b_1 < a_2 < b_2 < ... < a_k < b_k$, then for t "small" we have $M^t = (a_1^t, b_1^t) \cup (a_2^t, b_2^t) \cup ... \cup (a_k^t, b_k^t)$, where a^t, b^t are defined by rule 4) above. There exists a first moment $t_0 > 0$ such that two intervals meet, i.e. $b_s^{t_0} = a_{s+1}^{t_0}$. In this case, we define the set

$$N = (a_1^{t_0}, b_1^{t_0}) \cup (a_2^{t_0}, b_2^{t_0}) \cup ... \cup (a_s^{t_0}, b_{s+1}^{t_0}) ... \cup (a_k^{t_0}, b_k^{t_0})$$

as union of $k - 1$ intervals. For $t > 0$ the set M^{t_0+t} is defined as N^t up to
the moment when two intervals of N^t meet. Then, the same procedure is
continued. Since at each step the number of intervals diminishes by one, at
some moment we get only one interval. An arbitrary open set is decomposed
in a countable union of intervals, and the symmetrization is defined by interior approximation. Any measurable set is approached by open sets, the
symmetrization being defined by exterior approximation.

For $N \geq 2$, the CSS is defined with respect to a hyperplane. For example,
let us suppose that $\mathcal{H}$ is the hyperplane defined by $\mathcal{H} = \{x_N = 0\}$. If $M \subseteq$
$\mathbb{R}^N$ is a polyhedron with all its faces parallel to the coordinate hyperplanes,
then by definition

$$M^t = \{\bigcup (M \cap l_{(x',0)})^t : x' \in \mathbb{R}^{N-1}\}.$$

Here $x = (x', 0)$ and l_x denotes the line orthogonal to $\mathcal{H}$ passing through the
point x; the set $(M \cap l_{(x',0)})^t$ is the one dimensional continuous symmetrization
of $M \cap l_{(x',0)}$. For an open set, the CSS is defined by interior approximation
with a sequence of polyhedra and for a measurable set the CSS is defined by
exterior approximation with open sets.

For a quasi-open set A, the set A^t is defined in the following way: consider
a decreasing sequence of open sets $(A_n)_{n \in N}$ with $\operatorname{Cap}(A_n \setminus A) \to 0$ and $A \subseteq$
$A_n \subseteq B$. For any $t \in [0, 1]$ the set A_n^t is well defined, and by monotonicity
we define $A_n^t \supseteq A_{n+1}^t$. Then $(A_n^t)_{n \in N}$ is γ-convergent and

$$A^t = \gamma - \lim_{n \to \infty} A_n^t.$$

For any positive measurable function u we define the continuous Steiner
symmetrization of u by symmetrizing its level sets:

$$\forall s > 0 \ \{u^t > s\} := \{u > s\}^t.$$

In [32] it is proved the following.

Theorem 6.2.1 *Let* $u \in H^1(\mathbb{R}^N)$, $u \geq 0$. *Then* $u^t \in H^1(\mathbb{R}^N)$, $\|u\|_{L^2} =$
$\|u^t\|_{L^2}$ *and* $\|u\|_{H^1} \geq \|u^t\|_{H^1}$. *Moreover, if* Ω *is an open set and* $u \in$
$H_0^1(\Omega)$, $u \geq 0$ *then* $u^t \in H_0^1(\Omega^t)$.

For other properties concerning the continuous Steiner symmetrization
we refer to [32].

We recall from [44] some useful results (without proofs). Consider a measurable set A and a hyper-plane $\mathcal{H} \subseteq \mathbb{R}^N$. For $t \in [0,1]$ denote by A^t the Steiner symmetrization of A at time t in the orthogonal direction to $\mathcal{H}$.

Proposition 6.2.2 *For every bounded quasi-open set $A \subseteq \mathbb{R}^N$ and every positive integer i the mappings $t \mapsto \lambda_i(A^t)$, are lower semicontinuous on the left and upper semicontinuous on the right.*

For a compact set $K \in \mathbb{R}^N$ it is known the existence of a sequence of hyper-planes $(\mathcal{H}_n)_{n \in N}$ such that denoting $K_0 = K$ and K_n the symmetrization of K_{n-1} with respect to $\mathcal{H}_n$ we have that $m(K_n \Delta K^{\#}) \to 0$ (generally by $C^{\#}$ we denote the closed ball of measure $m(C)$, see [29]). If the convergence in measure is replaced by the Hausdorff convergence, a similar type of result can be found in Federer, see [116].

For quasi-open sets we can formulate the following.

Proposition 6.2.3 *Let A be a bounded quasi-open set of $\mathbb{R}^N$. There exists a sequence of Steiner symmetrizations of A, denoted by $(A_n)_{n \in N}$ such that $m(A_n \setminus A^{\#}) \to 0$ for $n \to \infty$.*

Proof This result appears to be weaker than the similar one for compact sets, but nevertheless it is still sufficient for our forthcoming purposes.

Suppose first that A is open. Consider $K_1 \subset\subset A$ such that $m(A \setminus K_1) \leq \epsilon_1/2$. We perform a finite number of Steiner symmetrizations given by the result of [29] for K_1 such that

$$m\left((K_1)_{n_1} \Delta K_1^{\#}\right) \leq \frac{\epsilon_1}{2}.$$

Then, by the monotonicity

$$m\left(A_{n_1} \setminus A^{\#}\right) \leq \frac{\epsilon_1}{2} + \frac{\epsilon_1}{2} = \epsilon_1.$$

Choosing now another set $K_2 \subset\subset A_{n_1}$ with $m(A_{n_1} \setminus K_2) \leq \epsilon_2/2$ we continue the process and obtain

$$m\left((K_2)_{n_2} \setminus K_2^{\#}\right) \leq \frac{\epsilon_2}{2}$$

and so on. Choosing a sequence $\epsilon_n \to 0$ we conclude the proof in the case of open sets.

If A is quasi-open, consider a sequence of bounded open sets $(C_r)_{r \in \mathbb{N}}$ such that

$$A \subseteq C_{r+1} \subseteq C_r$$

and $\text{Cap}(C_r \setminus A) \to 0$, and we apply the previous result for C_r. We make a finite number of symmetrizations to C_1 such that

$$m\left((C_1)_{n_1} \setminus C_1^{\#}\right) \leq \epsilon_1.$$

Then $m(A_{n_1} \setminus C_1^{\#}) \leq \epsilon_1$. Making now a finite number of symmetrizations for C_2 we get $m((C_2)_{n_2} \setminus C_2^{\#}) \leq \epsilon_2$, and so on. Finally we get $m(A_n \setminus A^{\#}) \to 0$, since $m(C_n^{\#} \Delta A^{\#}) \to 0$. ∎

Corollary 6.2.4 *For every bounded quasi-open set $A \subseteq \mathbb{R}^N$ there exists a sequence $(A_n)_{n \in \mathbb{N}}$ of successive Steiner symmetrizations of A such that any weak γ-limit point of $(A_n)_{n \in \mathbb{N}}$ is contained in $A^{\#}$.*

Proof Indeed, from the previous proposition we have $m(A_n \setminus A^{\#}) \to 0$. If U is the weak γ-limit of $\{A_{n_k}\}$, then $w_{n_k} \rightharpoonup w$ weakly in $H_0^1(B)$ and $U \Rightarrow \{w > 0\}$. Since $m(A_n \setminus A^{\#}) \to 0$, and $w_{n_k} \to w$ in $L^2(B)$ we get $w = 0$ a.e. on $\mathbb{R}^N \setminus A^{\#}$, hence $w \in H_0^1(A^{\#})$, which means $U \subseteq A^{\#}$. ∎

Corollary 6.2.5 *For the sequence $(A_n)_{n \in \mathbb{N}}$ given by Corollary 6.2.4 we have*

$$\lambda_k(A^{\#}) \leq \liminf_{n \to \infty} \lambda_k(A_n).$$

6.3 The case of the first two eigenvalues of the Laplace operator

Let $B \subseteq \mathbb{R}^N$ be a fixed ball and let $c > 0$ be a positive number. We denote by

$$\mathcal{A}_c(B) = \{A \subseteq B : \quad A \quad \text{quasi-open}, \ m(A) \leq c\}$$

the family of all quasi-open subsets of B having Lebesgue measure less than or equal to c and by $s : \mathcal{A}_c(B) \to \mathbb{R}^2$ the *spot* function defined by

$$s(A) = (\lambda_1(A), \lambda_2(A)),$$

where $\lambda_1(A), \lambda_2(A)$ are the first two eigenvalues (counted with their multi-plicities) of the Laplace operator $-\Delta$ on the Sobolev space $H_0^1(A)$.

The purpose of this section is to prove that the range of s is closed in $\mathbb{R}^2$, if for a given c, the ball B is large enough. This will immediately imply the existence of a solution for problems of the form

$$\min \left\{ \Phi(\lambda_1(A), \lambda_2(A)) : \quad A \in \mathcal{A}_c(B) \right\}$$

for a large class of cost functions Φ (in particular no monotonicity will be required).

Let us denote by $E = s(\mathcal{A}_c(B))$ the image of s in $\mathbb{R}^2$. The set E is conical with respect to the origin, that is $(tx, ty) \in E$ whenever $(x, y) \in E$ and $t \geq 1$. This is easily seen by considering the homothetical sets $A_t = A/\sqrt{t}$, where A is such that $s(A) = (x, y)$. Moreover, following the results of Faber, Krahn, Ashbaugh-Benguria, one has already an idea where the set E lies.

For a numerical study of the set E in the case $N = 2$ we refer the inter-ested reader to the paper by Wolf and Keller, see [184] where the following picture for E is obtained:

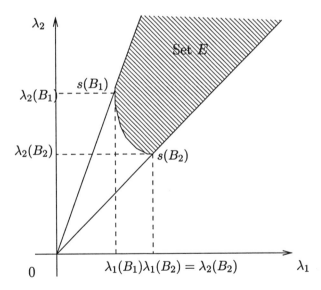

Figure 6.1: The set E for $N = 2$ and $c = 1$

Unfortunately, we are not able to prove the convexity of the set E, which the picture above seems to show; this would imply the closure result quite straightforward. However, we can prove that E is convex horizontally and vertically, and this is enough to imply that it is closed.

If we denote by φ_1, φ_2 the eigenfunctions corresponding to $\lambda_1(A)$ and $\lambda_2(A)$ then we have

$$\varphi_1 \in H_0^1(A) \quad \text{and} \quad -\Delta\varphi_1 = \lambda_1(A)\varphi_1 \quad \text{in} \quad H_0^1(A),$$
$$\varphi_2 \in H_0^1(A) \quad \text{and} \quad -\Delta\varphi_2 = \lambda_2(A)\varphi_2 \quad \text{in} \quad H_0^1(A).$$

We shall generally use the following notation: $A_1 = \{\varphi_1 > 0\}$, $A_2 = \{\varphi_2 \neq 0\}$, where we denote here by φ_1 and φ_2 the quasi-continuous representatives of the corresponding eigenfunctions.

Lemma 6.3.1 *Let A be a quasi-open set such that $\lambda_1(A) < \lambda_2(A)$. Then the fine interior of A_1 is finely connected and there are two possibilities: either $A_2 \subseteq A_1$ or $\mathrm{Cap}(A_1 \cap A_2) = 0$ for a convenient second eigenfunction φ_2.*

Proof If A is open the result is immediate. If A is quasi-open, the proof is similar and based on Lemma 4.1.3 and the following assertion (see [121]): any positive superharmonic function on a finely open and connected set is either strictly positive or equal to zero. Particularly, this will be the case of the first eigenfunction.

Indeed, if A_1 is not finely connected (we denoted here by A_1 its fine interior) then it can be decomposed in a union of disjoint finely connected components $(C_i)_{i \in I}$ and since $\varphi_{1|C_i} \in H_0^1(C_i) \subseteq H_0^1(A)$ we have that

$$\forall i \in I \quad \frac{\int_{C_i} |\nabla\varphi_{1|C_i}|^2 dx}{\int_{C_i} |\varphi_{1|C_i}|^2 dx} = \lambda_1(A)$$

So, if I contains at least two indices this would mean that $\lambda_1(A)$ is at least double, since we have two independent eigenfunctions (defined by the restriction of φ_1 on each set). Therefore A_1 has only one finely connected component.

Suppose now that $\mathrm{Cap}(A_1 \cap A_2) \neq 0$. Decomposing $A_2 = \cup_{i \in I} C_i'$, C_i' being finely connected, then for any component for which $\mathrm{Cap}(A_1 \cap C_i') \neq 0$ we have $C_i' \subseteq A_1$; indeed, otherwise $C_i' \cup A_1$ would be finely connected and φ_1 could not vanish on $C_i' \setminus A_1$. So the finely connected components of A_2 are of two types $C_i' \subseteq A_1$ and $\mathrm{Cap}(C_j' \cap A_1) = 0$. In this case we can see that $\varphi_{2|\cup C_i'}$ and $\varphi_{2|\cup C_j'}$ are both orthogonal to φ_1 and they are still second eigenfunctions. Then A_2 can be chosen as $\cup C_i'$ or $\cup C_j'$. ∎

Let $c > 0$ be given and let $B \subseteq \mathbb{R}^N$ be a ball containing two disjoint balls of mass $c/2$. We shall prove the following result.

Theorem 6.3.2 *The set E is closed in $\mathbb{R}^2$.*

The proof of the theorem above is based on the following lemma.

Lemma 6.3.3 *If the set E is convex on the vertical and horizontal directions, then E is closed in $\mathbb{R}^2$.*

Proof Consider $(x, y) \in \bar{E}$. There exists a sequence of sets $(A_n)_{n \in \mathbb{N}} \subseteq \mathcal{A}_c(B)$ such that $s(A_n) \to (x, y)$. From the weak γ-compactness of the set $\mathcal{A}_c(B)$, for a subsequence still denoted with the same indices we can write $A_n \to A$ in the weak γ-sense. Then $A \in \mathcal{A}_c(B)$ and since the eigenvalues of the Laplacian are weakly γ-lower semi-continuous we get $\lambda_1(A) \leq \liminf_{n \to \infty} \lambda_1(A_n) = x$ and $\lambda_2(A) \leq \liminf_{n \to \infty} \lambda_2(A_n) = y$. From the vertical convexity of E, the vertical segment joining $s(A)$ with the half line d_1 is contained in E. If $y < \lambda_2(B_1)$ we can find the point $(\lambda_1(A), y)$ on this segment and using now the horizontal convexity, the segment joining $(\lambda_1(A), y)$ to d_2 is in E. But this segment contains the point (x, y) since $\lambda_1(A) \leq x$.
If $y \geq \lambda_2(B_1)$, then the horizontal convexity gives directly $(x, y) \in E$. ∎

Following Lemma 6.3.3 it suffices to prove the convexity of E on vertical and horizontal directions. For this purpose, we split the proof in two steps :

Step 1 A is convex on horizontal lines, namely if $A \in \mathcal{A}_c(B)$ then the segment joining $(\lambda_1(A), \lambda_2(A))$ to $(\lambda_2(A), \lambda_2(A))$ is contained in E.

Step 2 A is convex on vertical lines, namely if $A \in \mathcal{A}_c(B)$ then the segment joining $(\lambda_1(A), \lambda_2(A))$ to $(\lambda_1(A), \frac{\lambda_2(B_1)}{\lambda_1(B_1)}\lambda_1(A))$ is contained in E.

In the proof of step 2, the idea is to make a sequence of CSS to transform a given quasi-open set $A \in \mathcal{A}_c(B)$ into a ball. Here, one can see that the choice of B is important since if $A \in \mathcal{A}_c(B)$ for any hyper-plane $\mathcal{H}$ we still have $A^t \in \mathcal{A}_c(B)$.

We proceed now with the proofs of steps 1 and 2. It is convenient to indicate by d_1 the half line $\{ts(B_1) : t \geq 1\}$ and by d_2 the half line $\{ts(B_2) : t \geq 1\} = \{(x, x) \in \mathbb{R}^2 : x \geq \lambda_1(B_2)\}$.

We give first a general result which establishes the existence of a γ-continuous and decreasing homotopy between two quasi-open sets $A_1 \subseteq A_0$.

Proposition 6.3.4 *Let $A_1 \subseteq A_0$ be two quasi-open sets. There exists a decreasing homotopy from A_0 to A_1 which is γ-continuous, namely there exists a γ-continuous mapping $h : [0,1] \to \mathcal{A}(\mathbb{R}^N)$ such that for $t_1 < t_2$, $h(t_1) \supseteq h(t_2)$ and $h(0) = A_0$, $h(1) = A_1$.*

Proof Denote by K a closed cube containing A_0. We shall divide the cube in 2^N equal closed cubes $K_0, \ldots, K_{2^N-1}$; each cube K_i is analogously divided in 2^N closed cubes $K_{i0}, \ldots, K_{i2^N-1}$, and so on. Then to each real number $t \in [0,1]$ written in the 2^N-basis by $0, \alpha_1, \alpha_2, \ldots$ we associate the set

$$\Lambda_t = (A_0 \setminus F_t) \cup A_1$$

where

$$F_t = \cup_{n=1}^\infty \cup_{i=0}^{\alpha_n-1} K_{\alpha_1 \ldots \alpha_{n-1} i}.$$

Remark first that Λ_t is quasi-open since F_t is quasi-closed. Indeed, let's denote by

$$F_{t,k} = \cup_{n=1}^k \cup_{i=0}^{\alpha_n-1} K_{\alpha_1 \ldots \alpha_{n-1} i}$$

the closed set consisting of the first k-blocks of F_t. Set also

$$\Lambda_{t,k} = (A_0 \setminus F_{t,k}) \cup A_1$$

which is obviously quasi-open, and remark that

$$\cap_{k \geq 1} \Lambda_{t,k} = \Lambda_t.$$

Since $\mathrm{Cap}(\Lambda_{t,k} \setminus \Lambda_t) \to 0$ for $k \to \infty$ we get that Λ_t is quasi-open. Moreover, the mapping $t \to \Lambda_t$ is continuous in capacity. Indeed, fix $t \in [0,1]$ and consider $t_n \to t$. We have to distinguish two situations: either t has an infinite number of digits and is not finishing with $aa...aa...$, or t has a finite number of digits or finishes with $aa...aa...$ (by a it is denoted the greatest digit in the basis 2^N, namely $a = 2^N - 1$). In the first case, if $t_n \to t$ then for every $k \in \mathbb{N}$ there exists $n_k \in \mathbb{N}$ such that for every $n \geq n_k$ the numbers t_n and t have the same first k digits. In this case

$$\mathrm{Cap}(A^t \Delta A^{t_n}) \leq \mathrm{Cap}(K_{\alpha_1 \ldots \alpha_k})$$

and we derive the continuity in capacity.

If t has a finite number of digits $t = 0.\alpha_1\alpha_2...\alpha_k$, then t written as

$$t = 0.\alpha_1\alpha_2...\alpha_k 0000...$$

is identified with

$$t' = 0.\alpha_1\alpha_2...(\alpha_k - 1)aaaa...$$

The difference between A^t and $A^{t'}$ is a point, hence of zero capacity. Consider $t_n \to t$. If $t_n \geq t$ the first k digits of t_n and t coincide for $n \geq n_k$. If $t_n < t$ then the first k digits of t_n and t' coincide for $n \geq n_k$ and the conclusion follows.

Since the mapping $t \mapsto \Lambda_t$ is obviously decreasing and γ-continuous, to achieve the proof it is enough to take

$$h(t) = \Lambda_t.$$

∎

Proof of step 1 : Let $A \in \mathcal{A}_c(B)$. If there exists a subset A^* of A, such that $\lambda_1(A^*) = \lambda_2(A^*) = \lambda_2(A)$, then one can apply directly Proposition 3.1, and step 1 is proved since there exists a decreasing and γ-continuous homotopy from A to A^*. Since $\lambda_2(A) = \lambda_2(A^*)$ then by monotonicity $\lambda_2(\Lambda_t) = \lambda_2(A)$. Since the first eigenvalue is γ-continuous, for each $\alpha \in [\lambda_1(A), \lambda_2(A)]$ there exists some t_α such that $\lambda_1(\Lambda_{t_\alpha}) = \alpha$.

Let's prove now the existence of the set A^*. If $\lambda_1(A) = \lambda_2(A)$, there is nothing to prove. Hence we suppose $\lambda_1(A) < \lambda_2(A)$ and from what we have seen in the previous section we have two possibilities: either $A_2 = \{\varphi_2 \neq 0\} \subseteq \{\varphi_1 > 0\} = A_1$ and in this case $A^* = A_2$, or $Cap(A_1 \cap A_2) = 0$ and denoting by P_t the open half space

$$P_t = \{(x_1, ..., x_N) \in \mathbb{R}^N | t < x_1\}$$

there exists some $t_0 \in \mathbb{R}$ such that $\lambda_1(A_1 \cap P_{t_0}) = \lambda_2(A)$. Choosing

$$A^* = (A_1 \cap P_{t_0}) \cup A_2$$

the conclusion follows. ∎

Proof of Step 2 : Let's consider $A \in \mathcal{A}_c(B)$ and denote by $A^\#$ the closed ball of measure $m(A)$. The idea to prove the convexity in the vertical

direction is to make a sequence of continuous Steiner symmetrizations trans-
forming A, such that $m(A_n \setminus A^\#) \to 0$, and to use the horizontal convexity.
If $\lambda_2(A) \geq \lambda_2(A^\#)$ then the segment

$$\left\{ (\lambda_1(A), \gamma) \; : \quad \text{for} \quad \gamma \in [\lambda_2(A), \lambda_1(A)\frac{\lambda_2(B_1)}{\lambda_1(B_1)}] \right\}$$

is contained in E. This follows immediately from the convexity on the horizontal lines since all the half line supported by d_1 and having B_1 as extreme
point is in E.

So let's suppose $\lambda_2(A) < \lambda_2(A^\#)$, and choose $\alpha \in]\lambda_2(A), \lambda_2(A^\#)[$. We
intend to prove that $(\lambda_1(A), \alpha) \in E$. We use Corollary 6.2.4 and we find a
sequence of continuous Steiner symmetrizations $(A_n)_{n \in \mathbb{N}}$ such that

$$\liminf_{n \to \infty} \lambda_2(A_n) \geq \lambda_2(A^\#).$$

In order to underline the evolution of the set A "towards" the ball, we say
that the CSS from A_n to A_{n+1} is parametrized by $t \in [n, n+1]$, by simple
translation of the interval $[0, 1]$. In this way we can define the set A^t for
every $t \geq 0$, and the set A_n can also be written as A^n.

On the other hand, $\lambda_1(A_n) \leq \lambda_1(A)$. There exists some $n_0 \in N$ such that
$\lambda_2(A_{n_0}) \geq \alpha$ and denote

$$t^* = \sup \left\{ t \in [0, n_0] \; : \quad \lambda_2(A^t) \leq \alpha \right\}.$$

From the upper semicontinuity on the right we have $\lambda_2(A^{t^*}) \geq \alpha$ and from the
lower semicontinuity on the left we get $\lambda_2(A^{t^*}) \leq \alpha$ which give $\lambda_2(A^{t^*}) = \alpha$.

Using now the convexity on the horizontal lines, the segment joining
$(\lambda_1(A^{t^*}), \alpha)$ with $(\alpha, \alpha) \in d_2$ is contained in E. But since $\lambda_1(A^{t^*}) \leq \lambda_1(A)$
the point $(\lambda_1(A), \alpha)$ belongs to E.   ∎

We remark that the previous result can be applied to prove the existence
of solutions for some classes of shape optimization problems, for which the
shape functional is not monotone with respect with the set inclusion (see
[64]). We can consider problems of the form

$$(6.4) \qquad \min \left\{ \Phi(\lambda_1(A), \lambda_2(A)) : \quad A \in \mathcal{A}_c(B) \right\}$$

where $\Phi : E \to \overline{\mathbb{R}}$ is lower semi-continuous and goes to $+\infty$ at infinity. This is the case for instance of

$$\Phi(x, y) = (x - \alpha)^2 + (y - \beta)^2$$

where (α, β) is any element in $\mathbb{R}^2$. Therefore by Theorem 6.3.2 the minimization problem (6.4) admits at least a solution.

Remark 6.3.5 An interesting question is to study the boundary of the set E. We can provide only some information on it. For a set $A \in \mathcal{A}_c(B)$ we denote by $\mathcal{R}_{\inf}(A)$ the rectangle

$$\mathcal{R}_{\inf}(A) = \{(x, y) \in \mathbb{R}^2 : x \leq \lambda_1(A), y \leq \lambda_2(A)\}.$$

For every $A \in \mathcal{A}_c(B)$, there exists a set $\tilde{A} \in \mathcal{A}_c(B)$ which is either finely connected or two balls, such that $s(\tilde{A}) \in \mathcal{R}_{\inf}(A)$.

Indeed, let us fix $A \in \mathcal{A}_c(B)$ and set $A_1 = \{\varphi_1 > 0\}$ and $A_2 = \{\varphi_2 \neq 0\}$. If $\lambda_1(A) = \lambda_2(A)$ the assertion is obvious since $s(A) \in d_2$. If $\lambda_1(A) < \lambda_2(A)$ there are two possibilities. If $A_2 \subseteq A_1$, then A_1 is finely connected and $s(A_1) \in \mathcal{R}_{\inf}(A)$. If $\mathrm{Cap}(A_2 \cap A_1) = 0$ we make the Schwarz rearrangements of A_1 and A_2 into the disjoint balls C_1 and C_2, and we get $s(C_1 \cup C_2) \in \mathcal{R}_{\inf}(A)$ and $C_1 \cup C_2 \in \mathcal{A}_c(B)$.

This proposition means that any A whose $s(A)$ is on $\partial E \setminus (d_1 \cup d_2)$ is either finely connected or two balls. An open question is to study if these sets are simply connected.

6.4 Unbounded design regions

The aim of this section is to give a variational method for proving global existence results for shape optimization problems depending on eigenvalues if the design region is $\mathbb{R}^N$. In order to develop the method presented in sections 6.1 and 6.3 to unbounded design regions, one has to relate the γ-convergence to the concentration-compactness principle, since the injection $H^1(\mathbb{R}^N) \hookrightarrow L^2(\mathbb{R}^N)$ fails to be compact, and the γ-convergence (which is still compact) does not yield the convergence of the eigenvalues. A detailed description of this subject can be found in [36].

The concentration-compactness principle describes the behavior in $L^2(\mathbb{R}^N)$ of a bounded sequence $(u_n)_{n \in \mathbb{N}}$ of $H^1(\mathbb{R}^N)$ (see [150]). More precisely, three situations may occur for a subsequence: compactness (maybe making some translations), vanishing or dichotomy. Given a sequence of open (or quasi-open) sets $(A_n)_{n\in\mathbb{N}}$ of $\mathbb{R}^N$, with uniformly bounded measure ($|A_n| \le c$, for all $n \in \mathbb{N}$), a natural question is to see whether *all* bounded sequences $(u_n)_{n\in\mathbb{N}}$ of $H^1(\mathbb{R}^N)$, such that u_n belongs to $H_0^1(A_n)$ for every $n \in \mathbb{N}$, have the *same* behavior in $L^2(\mathbb{R}^N)$ with respect to the concentration-compactness principle. This is particularly important from the point of view of shape optimization problems. It is of interest to know whether for a suitable sequence $(y_n)_{n\in\mathbb{N}} \subseteq \mathbb{R}^N$ the injection

(6.5)
$$\bigcup_{n\in\mathbb{N}} H_0^1(y_n + A_n) \hookrightarrow L^2(\mathbb{R}^N)$$

is compact, i.e. a bounded subset of $\bigcup_{n\in\mathbb{N}} H_0^1(y_n + A_n)$ for the $H^1(\mathbb{R}^N)$-norm is relatively compact in $L^2(\mathbb{R}^N)$ (by $y_n + A_n$, one denotes the translation of A_n by the vector y_n). Notice that if A is a quasi-open set of bounded measure, then $H_0^1(A)$ is compactly embedded in $L^2(A)$ (see [183]).

A related point of view on this question comes from the γ-convergence theory of sets in unbounded design regions. Following [19], [97], there exists a subsequence of $(A_n)_{n\in\mathbb{N}}$, still denoted with the same index, which γ-converges to a measure μ, i.e. for any bounded open set Ω, the sequence of functionals

$$F_n(u,\Omega) = \int_{\mathbb{R}^N} |\nabla u|^2 dx + \chi_{H_0^1(A_n \cap \Omega)}(u)$$

Γ-converges to

$$F(u,\Omega) = \int_{\mathbb{R}^N} |\nabla u|^2 dx + \int_{\mathbb{R}^N} u^2 d\mu + \chi_{H_0^1(\Omega)}(u)$$

in $L^2(\mathbb{R}^N)$.

If one denotes by R_{A_n} the resolvent operator of the Laplace equation with homogeneous Dirichlet boundary conditions respectively by R_μ the resolvent operator associated to the measure μ, a consequence of the γ-convergence is the following pointwise convergence of the resolvent operators:

(6.6)
$$\forall f \in L^2(\mathbb{R}^N) \quad R_{A_n}(f) \xrightarrow{L^2(\mathbb{R}^N)} R_\mu(f).$$

The γ-convergence may or may not involve the convergence of the resolvent operators in the operator norm; in bounded design regions the convergence in the operator norm is a consequence of the compact embedding

$H_0^1(D) \hookrightarrow L^2(D)$. We prove that injection (6.5) is compact, if and only if the convergence (6.6) is uniform in $L^2(\mathbb{R}^N)$. Consequently, the convergence of the resolvent operators will hold in the operator norm.

γ-convergence for unbounded domains. We recall in this paragraph some results on the γ-convergence of unbounded domains. Given a sequence of open sets $A_n \subseteq \mathbb{R}^N$, from Dal Maso-Mosco [97] (see also [19]) there exists a subsequence (still denoted with the same index) such that A_n γ-converges to a measure $\mu \in \mathcal{M}_0(\mathbb{R}^N)$, i.e. for any bounded open set Ω, the sequence of functionals

$$F_n(u, \Omega) = \int_{\mathbb{R}^N} |\nabla u|^2 dx + \chi_{H_0^1(A_n \cap \Omega)}(u)$$

Γ-converges to

$$F(u, \Omega) = \int_{\mathbb{R}^N} |\nabla u|^2 dx + \int_{\mathbb{R}^N} u^2 d\mu + \chi_{H_0^1(\Omega)}(u),$$

where $\chi_{H_0^1(A_n \cap \Omega)}(u) = 0$ if $u \in H_0^1(A_n \cap \Omega)$ and ∞ if not. Let us denote $F_n(u) = \int_{\mathbb{R}^N} |\nabla u|^2 dx + \chi_{H_0^1(A_n)}(u)$ and $F(u) = \int_{\mathbb{R}^N} |\nabla u|^2 dx + u^2 d\mu$.
   We suppose that $(A_n)_{n \in \mathbb{N}}$ is a sequence of quasi-open sets of uniformly bounded measure.

Lemma 6.4.1 *Let us suppose that A_n γ-converges to μ. Then for every sequence $u_n \rightharpoonup u$ weakly in $L^2(\mathbb{R}^N)$ we have $F(u) \leq \liminf\limits_{n \to \infty} F_n(u_n)$.*

Proof Let $u_n \rightharpoonup u$ weakly in $L^2(\mathbb{R}^N)$, and $\liminf\limits_{n \to \infty} F_n(u_n) < \infty$. We have then (after a renotation of the indices) $u_n \in H_0^1(A_n)$ and $\int_{\mathbb{R}^N} |\nabla u_n|^2 dx \leq M$. Moreover, we have $u_n \rightharpoonup u$ weakly in $H^1(\mathbb{R}^N)$.
   Consider some function $\rho_R \in C_0^\infty(\mathbb{R}^N)$, $\rho_R = 1$ on $B_{0,R}$, $\rho_R = 0$ on $\mathbb{R}^N \setminus B_{0,2R}$. Then $\rho_R u_n \rightharpoonup \rho_R u$ weakly in $H^1(\mathbb{R}^N)$, the convergence being strong in $L^2(\mathbb{R}^N)$ (from the compact injection $H_0^1(B_{0,2R}) \hookrightarrow L^2(\mathbb{R}^N)$).
   Taking $\Omega = B_{0,2R}$, the Γ-convergence definition gives

$$\liminf_{n \to \infty} \int_{\mathbb{R}^N} |\nabla \rho_R u_n|^2 dx \geq \int_{\mathbb{R}^N} |\nabla \rho_R u|^2 dx + \int |\rho_R u|^2 d\mu$$

and

$$\liminf_{n \to \infty} [\int_{\mathbb{R}^N} |\nabla u_n|^2 \rho_R^2 + |\rho_R u_n|^2 dx + 2 \int_{\mathbb{R}^N} \nabla u_n \nabla \rho_R u_n \rho_R dx + \int_{\mathbb{R}^N} |\nabla \rho_R|^2 u_n^2 dx]$$

$$= \liminf_{n \to \infty} \int_{\mathbb{R}^N} |\nabla u_n|^2 \rho_R^2 + |\rho_R u_n|^2 dx + 2 \int_{\mathbb{R}^N} \nabla u \nabla \rho_R u \rho_R dx + \int_{\mathbb{R}^N} |\nabla \rho_R|^2 u^2 dx,$$

which by hypothesis is greater or equal than

$$\int_{\mathbb{R}^N} |\nabla u|^2 \rho_R^2 + 2 \int_{\mathbb{R}^N} \nabla u \nabla \rho_R u \rho_R dx + \int_{\mathbb{R}^N} |\nabla \rho_R|^2 u^2 dx + \int_{\mathbb{R}^N} \rho_R^2 u^2 d\mu.$$

Hence

$$\liminf_{n \to \infty} \int_{\mathbb{R}^N} |\nabla u_n|^2 \rho_R^2 + |\rho_R u_n|^2 dx \geq \int_{\mathbb{R}^N} |\nabla u|^2 \rho_R^2 dx + \int_{\mathbb{R}^N} \rho_R^2 u^2 d\mu.$$

Making $R \to \infty$, we immediately get

$$\liminf_{n \to \infty} \int_{\mathbb{R}^N} |\nabla u_n|^2 + |u_n|^2 dx \geq \int_{\mathbb{R}^N} |\nabla u|^2 dx + \int_{\mathbb{R}^N} u^2 d\mu.$$

∎

Lemma 6.4.2 *Let us suppose that A_n γ-converges to μ. Then, for all $u \in L^2(\mathbb{R}^N)$ such that $F(u) < \infty$, there exists a sequence $u_n \in L^2(\mathbb{R}^N)$ strongly convergent in $L^2(\mathbb{R}^N)$ to u such that $F_n(u_n) \to F(u)$.*

Proof Let us consider $u \in L^2(\mathbb{R}^N)$ with $F(u) < \infty$. The sequence $\rho_R u$ converges in $L^2(\mathbb{R}^N)$ to u for $R \to \infty$. Moreover $F(\rho_R u) \to F(u)$. Indeed,

$$F(\rho_R u) = \int_{\mathbb{R}^N} |\nabla \rho_R u|^2 dx + \int_{\mathbb{R}^N} |\rho_R u|^2 d\mu$$

$$= \int_{\mathbb{R}^N} |\nabla u|^2 \rho_R^2 dx + 2 \int_{\mathbb{R}^N} \nabla u \nabla \rho_R u \rho_R dx + \int_{\mathbb{R}^N} |\nabla \rho_R|^2 u^2 dx + \int_{\mathbb{R}^N} |\rho_R u|^2 d\mu.$$

Making $R \to \infty$, from the Lebesgue dominate convergence theorem, we have

$$\int_{\mathbb{R}^N} |\nabla u|^2 \rho_R^2 dx \to \int_{\mathbb{R}^N} |\nabla u|^2 dx \quad \text{and} \quad \int_{\mathbb{R}^N} |\rho_R u|^2 d\mu \to \int_{\mathbb{R}^N} u^2 d\mu.$$

The functions ρ_R can be chosen such that $\|\nabla \rho_R\|_{L^\infty(\mathbb{R}^N)} \to 0$ (for example $\rho_R(x) = \rho_1(x/R)$). Then

$$\int_{\mathbb{R}^N} \nabla u \nabla \rho_R u \rho_R dx \to 0 \quad \text{and} \quad \int_{\mathbb{R}^N} |\nabla \rho_R|^2 u^2 dx \to 0 \quad \text{for } R \to \infty.$$

But, for all $R > 0$ there exists a sequence $u_n^R \longrightarrow \rho_R u$ strongly in $L^2(\mathbb{R}^N)$ such that $F_n(u_n^R) = F_n(u_n^R, B_{0,2R}) \to F(\rho_R u, B_{0,2R}) = F(\rho_R u)$. Then by a diagonal construction we find a sequence $(u_n)_{n \in \mathbb{N}}$ such that $u_n \longrightarrow u$ strongly in $L^2(\mathbb{R}^N)$ and $F_n(u_n) \longrightarrow F(u)$. ∎

Proposition 6.4.3 *Let us suppose that A_n γ-converges to μ. Then F_n Γ-converges to F both in $L^2(\mathbb{R}^N)$-strong and in $L^2(\mathbb{R}^N)$-weak.*

Proof One has to verify the two conditions of the Γ-convergence. These follow directly from Lemmas 6.4.1 and 6.4.2. ∎

Theorem 6.4.4 *Let us suppose that A_n is a sequence of quasi-open sets of uniformly bounded measure which γ-converges to μ, and moreover $w_{A_n} \longrightarrow w$ strongly in $L^2(\mathbb{R}^N)$. Denoting $A = \{w > 0\}$, the sequence of functionals*

$$G_n(u) = \int_{\mathbb{R}^N} |\nabla u|^2 dx + \chi_{H_0^1(A_n)}(u) - 2 \int_{A_n} u dx$$

Γ-converges in $L^2(\mathbb{R}^N)$ to

$$G(u) = \int_{\mathbb{R}^N} |\nabla u|^2 dx + \int_{\mathbb{R}^N} u^2 d\mu - 2 \int_A u dx.$$

Moreover, w is a minimizer for G.

Proof The proof is an immediate consequence of Proposition 6.4.3 and of the following fact: let $u_n \in H_0^1(A_n)$ such that $u_n \longrightarrow u$ strongly in $L^2(\mathbb{R}^N)$. Then $\int_{A_n} u_n dx \to \int_A u dx$. Indeed, we remark firstly that $|A| \leq \liminf_{n \to \infty} |A_n|$. This follows from the pointwise convergence a.e. of a subsequence of w_{A_n} to w. Secondly, we have $A = A_\mu$ (up to a set of zero capacity), where A_μ is the regular set of the measure μ, i.e. the union of all finely open sets of finite μ measure. Indeed, from the previous proposition we have that $w \in L_\mu^2(\mathbb{R}^N)$, hence $\{w > 0\} \subseteq A_\mu$. On the other side, for any bounded set D, we have that $A_\mu \cap D$ coincides with the set $\{w_D > 0\}$, where w_D is the weak $H^1(\mathbb{R}^N)$ limit of the sequence $w_{A_n \cap D}$. By the maximum principle we have $0 \leq w_{A_n \cap D} \leq w_{A_n}$, hence $\{w_D > 0\} \subseteq \{w > 0\}$. Since this is true for any bounded set D, we get $A_\mu \subseteq \{w > 0\}$. Hence $u \in H_0^1(A)$.

Moreover,

$$\left| \int_{A_n} u_n dx - \int_A u dx \right| \leq \int_{A_n \cup A} |u_n - u| dx$$
$$\leq |A_n \cup A|^{\frac{1}{2}} \left(\int_{A_n \cup A} |u_n - u|^2 dx \right)^{\frac{1}{2}} \to 0.$$

■

Since $w_{A_n} \longrightarrow w$ strongly in $L^2(\mathbb{R}^N)$, it is clear that w is the minimizer of G, hence it satisfies the equation

$$-\Delta w + \mu w = 1 \quad \text{in} \quad H^1(\mathbb{R}^N) \cap L^2_\mu(\mathbb{R}^N).$$

Remark 6.4.5 If A_n is a sequence of quasi-open sets of uniformly bounded measure which γ-converges to μ, denoting $A = \{w > 0\}$ the sequence of functionals

$$H_n(u) = \int_{\mathbb{R}^N} |\nabla u|^2 dx + \chi_{H^1_0(A_n)}(u) - 2 \int_A u \, dx$$

Γ-converges in $L^2(\mathbb{R}^N)$-weak to

$$H(u) = \int_{\mathbb{R}^N} |\nabla u|^2 dx + \int_{\mathbb{R}^N} u^2 d\mu - 2 \int_A u \, dx.$$

This follows from Proposition 6.4.3 and the continuity of the mapping $u \to \int_A u \, dx$ in $L^2(\mathbb{R}^N)$-weak. Therefore the minimizer of H_n denoted $\tilde{w}_{A_n}$ converges weakly in $L^2(\mathbb{R}^N)$ to w.

Remark 6.4.6 Like in the uniform bounded case, if $(A_n)_{n \in \mathbb{N}}$ is a sequence of uniform bounded measure sets, such that $w_{A_n} \rightharpoonup w$ weakly in $L^2(\mathbb{R}^N)$, then A_n γ-converges to a measure μ and $w = w_\mu$.

Indeed, since for a subsequence there exists a measure μ such that A_{n_k} γ-converges to μ, we get from Proposition 6.4.3 that $F_{A_{n_k}}$ Γ-converges in L^2-weak to F_μ, hence w is a minimizer for F_μ. We get $w = w_\mu$, and from the uniqueness of the measure μ satisfying this equality, we get that the entire sequence γ-converges to μ.

Statement of the main results. We give two theorems. The first one proves that if $R_{A_n}(1)$ converges strongly in $L^2(\mathbb{R}^N)$ then any weakly convergent sequence $u_n \in H^1(\mathbb{R}^N)$, such that, $u_n \in H^1_0(A_n)$ is strongly convergent in $L^2(\mathbb{R}^N)$.

Theorem 6.4.7 *Let $(A_n)_{n \in \mathbb{N}}$ be a sequence of open (or quasi-open) sets of uniformly bounded measure. If $R_{A_n}(1) \longrightarrow w$ in $L^2(\mathbb{R}^N)$, then for any sequence $(u_n)_{n \in \mathbb{N}}$ such that $u_n \in H^1_0(A_n)$ and $u_n \rightharpoonup u$ weakly in $H^1(\mathbb{R}^N)$ we have $u_n \longrightarrow u$ in $L^2(\mathbb{R}^N)$, i.e. injection (6.5) is compact.*

The second result describes the behavior of the sequence of Sobolev spaces through the resolvent operators.

Theorem 6.4.8 *Let $(A_n)_{n \in \mathbb{N}}$ be a sequence of open (or quasi-open) sets of uniformly bounded measure. There exists a subsequence (still denoted with the same indices) such that one of the following situations occurs:*

Compactness: *There exists a sequence of vectors $(y_n)_{n \in \mathbb{N}} \subseteq \mathbb{R}^N$ and a positive Borel measure μ, vanishing on sets of zero capacity, such that $y_n + A_n$ γ-converges to the measure μ and $R_{y_n + A_n}$ converges in the uniform operator topology of $L^2(\mathbb{R}^N)$ to R_μ.*

Dichotomy: *There exists a sequence of subsets $\tilde{A}_n \subseteq A_n$, such that*

$$\|R_{A_n} - R_{\tilde{A}_n}\|_2 \to 0, \quad and \quad \tilde{A}_n = A_n^1 \cup A_n^2$$

with $d(A_n^1, A_n^2) \to \infty$ and $\liminf\limits_{n \to \infty} |A_n^i| > 0$ for $i = 1, 2$.

Notice that, contrary to the concentration-compactness principle for functions, the vanishing situation does not appear in this theorem. Vanishing is, roughly speaking, covered by compactness, since if vanishing occurs for the sequence $(R_{A_n}(1))$ we prove that $R_{A_n} \to 0$ in the norm operator of $\mathcal{L}(L^2(\mathbb{R}^N))$.

In this section we prove Theorems 6.4.7 and 6.4.8. The idea is to use the concentration-compactness principle (see [111], [139], [150]) for the sequence of functions $(w_{A_n})_{n \in \mathbb{N}}$ and to discuss separately every case.

The concentration-compactness principle: Let $(u_n)_{n \in \mathbb{N}}$ be a bounded sequence in $H^1(\mathbb{R}^N)$ with $\int_{\mathbb{R}^N} u_n^2 dx \to \lambda > 0$. There exists a subsequence $(n_k)_{k \in \mathbb{N}}$ satisfying one of the following three possibilities:

i) *(compactness)* there exists $y_k \in \mathbb{R}^N$ such that

$$(6.7) \qquad \forall \varepsilon > 0, \exists R < \infty, \int_{y_k + B_{0,R}} u_{n_k}^2 dx \geq \lambda - \varepsilon;$$

ii) *(vanishing)*

$$(6.8) \qquad \lim_{k \to \infty} \sup_{y \in \mathbb{R}^N} \int_{y + B_{0,R}} u_{n_k}^2 dx = 0, \quad \text{forall } R < \infty;$$

iii) *(dichotomy)* there exists $\alpha \in (0, \lambda)$, there exists $k_0 \geq 1$, u_k^1, u_k^2 bounded in $H^1(\mathbb{R}^N)$ satisfying for $k \geq k_0$:

(6.9)
$$\|u_{n_k} - (u_k^1 + u_k^2)\|_{L^2(\mathbb{R}^N)} \leq \delta(\varepsilon) \to 0 \quad \text{for } \varepsilon \to 0^+,$$
$$|\int_{\mathbb{R}^N} (u_k^1)^2 dx - \alpha| \leq \varepsilon \quad \text{and} \quad |\int_{\mathbb{R}^N} (u_k^2)^2 dx - (\lambda - \alpha)| \leq \varepsilon,$$
$$\text{dist}\,(\text{supp } u_k^1, \text{ supp } u_k^2) \to \infty \quad \text{for } k \to \infty,$$
$$\liminf_{n \to \infty} \int_{\mathbb{R}^N} [|\nabla u_{n_k}|^2 - |\nabla u_k^1|^2 - |\nabla u_k^2|^2] dx \geq 0.$$

In order to apply the concentration-compactness principle for the sequence $(w_{A_n})_{n \in \mathbb{N}}$, we prove a lemma showing that this sequence is bounded in $H^1(\mathbb{R}^N)$ by a constant depending only on $|A|$.

Lemma 6.4.9 *Let A be a quasi-open set finite measure. There exists a constant M which depends only on $|A|$ such that*

$$\|w_A\|_{H^1(\mathbb{R}^N)} \leq M.$$

Proof Taking w_A as test function in equation (4.32) for $f \equiv 1$ we get $\int_A |\nabla w_A|^2 dx = \int_A w_A dx$. Hence using the Poincaré inequality with the constant β which depends only on $|A|$ we get

$$\|w_A\|_{H^1(\mathbb{R}^N)}^2 \leq \beta^2 \int_A |\nabla w_A|^2 dx = \beta^2 \int_A w_A dx \leq \beta^2 |A|^{\frac{1}{2}} \|w_A\|_{H^1(\mathbb{R}^N)},$$

concluding the proof.   ∎

The key result to prove Theorem 6.4.7 is the following:

Lemma 6.4.10 *Let $(A_n)_{n \in \mathbb{N}}$ be a sequence of quasi-open sets of uniformly bounded measure which γ-converges to a measure μ, and suppose that $w_{A_n} \to w$ in $L^2(\mathbb{R}^N)$. Then for any sequence $v_n \in H_0^1(A_n)$ such that $v_n \rightharpoonup v$ weakly in $H^1(\mathbb{R}^N)$ we have*

$$\int_{\mathbb{R}^N} v_n dx \to \int_{\mathbb{R}^N} v dx.$$

Proof Let us denote by A the quasi-open set $\{w > 0\}$. Then

$$\int_{A_n} \nabla w_{A_n} \nabla v_n dx = \int_{A_n} v_n dx$$

and

$$\int_A \nabla w_A \nabla v dx + \int_A wv d\mu = \int_A v dx.$$

We have the following estimation

$$|\int_{\mathbb{R}^N} \nabla w_{A_n} \nabla v_n dx - \int_A \nabla w_A \nabla v dx + \int_A wv d\mu|$$

$$\leq |\int_{\mathbb{R}^N} (\nabla w_{A_n} - \nabla \tilde{w}_{A_n}) \nabla v_n dx|$$

$$+ |\int_{\mathbb{R}^N} \nabla \tilde{w}_{A_n} \nabla v_n dx - \int_A \nabla w_A \nabla v dx - \int_A wv d\mu|$$

where $\tilde{w}_{A_n}$ denotes the solution of

(6.10)
$$\begin{cases} -\Delta \tilde{w}_{A_n} = 1_{A_n \cap A} \\ \tilde{w}_{A_n} \in H_0^1(A_n). \end{cases}$$

Usual Γ-convergence arguments give that $\tilde{w}_{A_n}$ converges weakly in $H^1(\mathbb{R}^N)$ to w. But

$$|\int_{\mathbb{R}^N} \nabla \tilde{w}_{A_n} \nabla v_n dx - \int_A \nabla w_A \nabla v dx - \int_A wv d\mu|$$

$$= |\int_{\mathbb{R}^N} 1_{A_n \cap A} v_n dx - \int_{\mathbb{R}^N} 1_A v dx|$$

$$= |\int_{\mathbb{R}^N} 1_A v_n dx - \int_{\mathbb{R}^N} 1_A v dx| \to 0.$$

On the other side

$$|\int_{\mathbb{R}^N} (\nabla w_{A_n} - \nabla \tilde{w}_{A_n}) \nabla v_n dx| \leq (\int_{\mathbb{R}^N} |\nabla v_n|^2 dx)(\int_{\mathbb{R}^N} |\nabla w_{A_n} - \nabla \tilde{w}_{A_n}|^2 dx).$$

It remains to prove that $\int_{\mathbb{R}^N} |\nabla w_{A_n} - \nabla \tilde{w}_{A_n}|^2 dx \to 0$, since the other term of the product in the right hand side, is bounded. But

(6.11)
$$\int_{\mathbb{R}^N} |\nabla w_{A_n} - \nabla \tilde{w}_{A_n}|^2 dx$$
$$= \int_{\mathbb{R}^N} |\nabla w_{A_n}|^2 dx - 2 \int_{\mathbb{R}^N} \nabla w_{A_n} \nabla \tilde{w}_{A_n} dx + \int_{\mathbb{R}^N} |\nabla \tilde{w}_{A_n}|^2 dx$$
$$= \int_{\mathbb{R}^N} w_{A_n} dx - 2 \int_{\mathbb{R}^N} \tilde{w}_{A_n} dx + \int_{\mathbb{R}^N} \tilde{w}_{A_n} 1_A dx.$$

Since w_{A_n} converges strongly in $L^2(\mathbb{R}^N)$ to w and since w_{A_n} are uniformly bounded (from Lemma 6.4.9), we get $\int_{\mathbb{R}^N} w_{A_n} dx \to \int_{\mathbb{R}^N} w dx$. Of course, we used that $|A_n|, |A| \leq c$. If we prove that $\tilde{w}_{A_n}$ converges strongly in $L^2(\mathbb{R}^N)$ to w, we get that the expression in (6.11) goes to zero.

In order to prove the strong L^2-convergence of $\tilde{w}_{A_n}$ to w, let us denote by $\tilde{w}^r_{A_n}$ the weak solution of the following problem

(6.12)
$$\begin{cases} -\Delta \tilde{w}^r_{A_n} = 1_{A_n \cap A \cap B_{0,r}} \\ \tilde{w}^r_{A_n} \in H^1_0(A_n \cap B_{0,r}). \end{cases}$$

The maximum principle yields $\tilde{w}_{A_n} \geq \tilde{w}^r_{A_n}$. But

$$\int_{\mathbb{R}^N} |\tilde{w}^r_{A_n}|^2 dx \longrightarrow \int_{\mathbb{R}^N} |w^r|^2 dx,$$

where by w_r we denote the solution of

(6.13)
$$\begin{cases} -\Delta w^r + \mu w^r = 1_{A \cap B_{0,r}} \\ \tilde{w}^r \in H^1_0(B_{0,r}) \cap L^2_\mu(B_{0,r}). \end{cases}$$

Hence

$$\liminf_{n \to \infty} \int_{\mathbb{R}^N} |\tilde{w}_{A_n}|^2 dx \geq \int_{\mathbb{R}^N} |w^r|^2 dx.$$

Making $r \to \infty$ and using the fact that $w^r \to w$ strongly in $L^2(\mathbb{R}^N)$ we get

$$\limsup_{n \to \infty} \int_{\mathbb{R}^N} |\tilde{w}_{A_n}|^2 dx \geq \int_{\mathbb{R}^N} w^2 dx.$$

On the other side $w_{A_n} \geq \tilde{w}_{A_n} \geq 0$, and $w_{A_n} \longrightarrow w$ strongly in $L^2(\mathbb{R}^N)$ hence

$$\lim_{n \to \infty} \int_{\mathbb{R}^N} |\tilde{w}_{A_n}|^2 dx = \int_{\mathbb{R}^N} w^2 dx,$$

which gives $\tilde{w}_{A_n} \longrightarrow w$ strongly in $L^2(\mathbb{R}^N)$.   ■

Proof [of Theorem 6.4.7] Let us suppose $u_n \in H_0^1(A_n)$ and $u_n \rightharpoonup u$ weakly in $H^1(\mathbb{R}^N)$. We prove that $\|u_n - u\|_{L^2(\mathbb{R}^N)} \to 0$, for $n \to \infty$. Performing the Fourier transform, we have

$$
\begin{aligned}
\|u_n - u\|_{L^2(\mathbb{R}^N)}^2 &= \int_{\mathbb{R}^N} |\widehat{u_n(y)} - u(y)|^2 dy \\
&= \int_{|y|>R} (1 + |y|^2)^{-1}(1 + |y|^2)|\widehat{u_n(y)} - u(y)|^2 dy \\
&\quad + \int_{|y|<R} |\widehat{u_n(y)} - u(y)|^2 dy \\
&\leq \frac{1}{1 + R^2}\|u_n - u\|_{H^1(\mathbb{R}^N)}^2 + \int_{|y|<R} |\widehat{u_n(y)} - u(y)|^2 dy.
\end{aligned}
$$

Let us fix $\varepsilon > 0$. Since the sequence $(u_n)_{n \in \mathbb{N}}$ is bounded in $H^1(\mathbb{R}^N)$, there exists $R > 0$, such that $\frac{1}{1+R^2}\|u_n - u\|_{H^1(\mathbb{R}^N)}^2 \leq \frac{\varepsilon}{2}$ for any $n \in \mathbb{N}$. With a fixed R, it remains to prove the existence of $n = n(R, \varepsilon) \in \mathbb{N}$ such that for all $n \geq n(R, \varepsilon)$ we have

$$
\int_{|y|<R} |\widehat{u_n(y)} - u(y)|^2 dy \leq \frac{\varepsilon}{2}.
$$

In fact, it is sufficient to prove that $\int_{|y|<R} |\widehat{u_n(y)} - u(y)|^2 dy \to 0$, for which we use the Lebesgue dominated convergence theorem. Fix $y \in B_{0,R}$ and consider the function $g_y(x) = e^{2i\pi\langle x,y \rangle}$. Since $u_n \in H_0^1(A_n)$ and $|A_n| \leq c$ we have $u_n g \in H_0^1(A^n)$. By definition, we have

$$
\hat{u}_n(y) = \int_{A_n} u_n(x) g_y(x) dx
$$

and

$$
\hat{u}(y) = \int_A u(x) g_y(x) dx.
$$

Therefore $|\hat{u}_n(y) - \hat{u}(y)| \to 0$, if we prove that

(6.14)
$$
\int_{A_n} u_n(x) g_y(x) dx \to \int_A u(x) g_y(x) dx.
$$

On the other hand, we have $u_n g_y \rightharpoonup u g_y$ weakly in $H^1(\mathbb{R}^N)$. This does not imply relation (6.14) immediately, since $1_{\mathbb{R}^N} \notin L^2(\mathbb{R}^N)$, but is a consequence of Lemma 6.4.10. Hence applying the Lebesgue dominate convergence theorem in $B_{0,R}$ we conclude the proof. ∎

Let now $(A_n)_{n\in\mathbb{N}}$ be a sequence of open (or quasi-open and non necessarily bounded) sets in $\mathbb{R}^N$ of uniformly bounded measure (say $|A_n| \leq c$). In order to prove Theorem 6.4.8, we apply the concentration-compactness principle to the sequence $(w_{A_n})_{n\in\mathbb{N}}$ which is bounded in $H^1(\mathbb{R}^N)$. Without loss of generality, we can suppose that $\int_{\mathbb{R}^N} w_{A_n}^2\,dx \to \lambda \geq 0$. We study separately each situation. The compactness and the vanishing cases will give uniform convergence for the sequence of operators $(R_{A_n})_{n\in\mathbb{N}}$, while dichotomy of $(w_{A_n})_{n\in\mathbb{N}}$ will give a dichotomy like behavior for $(R_{A_n})_{n\in\mathbb{N}}$.

Compactness Let us suppose that for a subsequence (still denoted with the same indices) and some translations (again we renote $y_n + A_n$ by A_n) we have the L^2-strong convergence of the sequence $(w_{A_n})_{n\in\mathbb{N}}$. Following γ-convergence arguments, if $w_{A_n} \longrightarrow w$ in $L^2(\mathbb{R}^N)$, then A_n γ-converges to a measure μ and $w = w_\mu$. Moreover, we have the following.

Proposition 6.4.11 *Let $(A_n)_{n\in\mathbb{N}}$ be a sequence of quasi-open sets of uniformly bounded measure and suppose that $w_{A_n} \longrightarrow w$ in $L^2(\mathbb{R}^N)$. Then R_{A_n} converges in $\mathcal{L}(L^2(\mathbb{R}^N))$ to R_μ.*

Proof We have to prove that

$$\lim_{n\to\infty} \sup_{\|f\|_{L^2(\mathbb{R}^N)}\leq 1} \|R_{A_n}(f) - R_\mu(f)\|_{L^2(\mathbb{R}^N)} = 0.$$

This is equivalent to

$$\lim_{n\to\infty} \|R_{A_n}(f_n) - R_\mu(f_n)\|_{L^2(\mathbb{R}^N)} = 0,$$

where $\|f_n\|_{L^2(\mathbb{R}^N)} \leq 1$. It is sufficient to consider a subsequence, denoted with the same indices, which weakly converges in $L^2(\mathbb{R}^N)$ to a function f. Hence

$$\limsup_{n\to\infty} \|R_{A_n}(f_n) - R_\mu(f_n)\|_{L^2(\mathbb{R}^N)}$$

$$= \limsup_{n\to\infty} \|R_{A_n}(f_n) - R_\mu(f) + R_\mu(f) - R_\mu(f_n)\|_{L^2(\mathbb{R}^N)}$$

$$\leq \limsup_{n\to\infty} \|R_{A_n}(f_n) - R_\mu(f)\|_{L^2(\mathbb{R}^N)} + \limsup_{n\to\infty} \|R_\mu(f) - R_\mu f_n\|_{L^2(\mathbb{R}^N)}.$$

But, from Remarks 6.4.6 and 6.4.5, we have $R_{A_n}(f_n) \rightharpoonup R_\mu(f)$ weakly in $H^1(\mathbb{R}^N)$. Applying Theorem 6.4.7, this convergence is strong in $L^2(\mathbb{R}^N)$. On the other side, $R_\mu(f_n) \rightharpoonup R_\mu(f)$ weakly in $H^1(\mathbb{R}^N)$. Denoting again by A the regular set of the measure μ (we have $|A| \leq c$), the compact injection $H_0^1(A) \hookrightarrow L^2(A)$ proves that this convergence is also strong in $L^2(\mathbb{R}^N)$. ∎

Corollary 6.4.12 *Under the hypotheses of proposition 6.4.11, we have $\lambda_k(A_n) \to \lambda_k(\mu)$, where by $\lambda_k(A_n)$ we denoted the k-th eigenvalue computed with its multiplicity of the Laplacian in $H_0^1(A_n)$.*

Proof It is a direct consequence of [109, Lemma XI.9.5]. ∎

Vanishing Let us suppose that $(w_{A_n})_{n \in \mathbb{N}}$ is in the vanishing case, i.e. for all $R > 0$

$$(6.15) \qquad \lim_{n \to \infty} \sup_{y \in \mathbb{R}^N} \int_{B_{y,R}} w_{A_n}^2 \, dx = 0.$$

We prove that $\lambda_1(A_n) \to \infty$ and hence any $H^1(\mathbb{R}^N)$-bounded sequence of elements $u_n \in H_0^1(A_n)$ converges strongly in $L^2(\mathbb{R}^N)$ to 0. This comes from the inequality $\|u_n\|_{L^2(A_n)} \leq \frac{1}{\lambda_1(A_n)} \|\nabla u_n\|_{L^2(A_n, \mathbb{R}^N)}$. In particular this will be the case of w_{A_n}, hence we find from the previous sub section that R_{A_n} converges to 0 in $\mathcal{L}(L^2(\mathbb{R}^N))$.

Proposition 6.4.13 *Let us suppose that $w_{A_n} \rightharpoonup w$ weakly in $H^1(\mathbb{R}^N)$ such that (6.15) holds. Then $\lambda_1(A_n) \to \infty$ and $\|R_{A_n}\|_2 \to 0$.*

Proof We use a result of Lieb from [149], namely that for any $\varepsilon > 0$, there exists some $R > 0$ and $y_n \in \mathbb{R}^N$ such that

$$(6.16) \qquad \lambda_1(A_n \cap B_{y_n, R}) \leq \lambda_1(A_n) + \varepsilon.$$

The maximum principle yields $w_{A_n} \geq w_{A_n \cap B_{y_n, R}} \geq 0$, hence relation (6.15) gives

$$\lim_{n \to \infty} \int_{A_n \cap B_{y_n, R}} w_{A_n \cap B_{y_n, R}}^2 \, dx = 0.$$

Translating A_n by the vector $-y_n$, we can suppose (maybe extracting a subsequence, still denoted with the same indices), that the sequence of sets $(-y_n + A_n) \cap B_{0,R}$ γ-converges to the empty set, which implies

$$\lambda_1((-y_n + A_n) \cap B_{0,R}) \to +\infty.$$

Hence, relation (6.16) gives that $\lambda_1(A_n) \to +\infty$ (see for example [64]). Then $w_{A_n} \longrightarrow 0$ in $L^2(\mathbb{R}^N)$ and from Proposition 6.4.11 we get $R_{A_n} \longrightarrow 0$ in $\mathcal{L}(L^2(\mathbb{R}^N))$. ∎

Dichotomy Supposing that $(w_{A_n})_{n \in \mathbb{N}}$ is in the dichotomy case, by a diagonal procedure we find a subsequence (still denoted with the same indices) such that there exists $\alpha > 0$ and $u_n^1, u_n^2 \in H^1(\mathbb{R}^N)$ with

(6.17) $$\|w_{A_n} - (u_n^1 + u_n^2)\|_{L^2(\mathbb{R}^N)} \to 0,$$

(6.18) $$\int_{\mathbb{R}^N} (u_n^1)^2 dx \to \alpha \quad \text{and} \quad \int_{\mathbb{R}^N} (u_n^2)^2 dx \to \lambda - \alpha,$$

(6.19) $$\text{dist}\,(\text{supp } u_n^1, \text{ supp } u_n^2) \to \infty,$$

(6.20) $$\liminf_{n \to \infty} \int_{\mathbb{R}^N} [|\nabla w_{A_n}|^2 - |\nabla u_n^1|^2 - |\nabla u_n^2|^2] dx \geq 0.$$

It is easy to see that u_n^1, u_n^2 can be chosen positive, and belonging to $H_0^1(A_n)$ (see the construction of u_n^1, u_n^2 in [150]). We define

$$\tilde{A}_n = A_n^1 \cup A_n^2 \quad \text{where} \quad A_n^1 = \{u_n^1 > 0\} \quad \text{and} \quad A_n^2 = \{u_n^2 > 0\},$$

which is a quasi-open set contained in A_n. We prove first that

(6.21) $$\|w_{A_n} - w_{\tilde{A}_n}\|_{H^1(\mathbb{R}^N)} \to 0.$$

Remark that $w_{\tilde{A}_n} = P_{H_0^1(\tilde{A}_n)} w_{A_n}$, where $P_{H_0^1(\tilde{A}_n)}$ denotes the orthogonal projection from $H_0^1(A_n)$ onto $H_0^1(\tilde{A}_n)$. Then

$$\int_{\mathbb{R}^N} |\nabla w_{A_n} - \nabla w_{\tilde{A}_n}|^2 dx \leq \int_{\mathbb{R}^N} |\nabla w_{A_n} - \nabla u_n^1 - \nabla u_n^2|^2 dx$$

$$= \int_{\mathbb{R}^N} |\nabla w_{A_n}|^2 dx - 2 \int_{\mathbb{R}^N} \nabla w_{A_n} \nabla (u_n^1 + u_n^2) dx + \int_{\mathbb{R}^N} |\nabla (u_n^1 + u_n^2)|^2 dx$$

$$= \int_{\mathbb{R}^N} w_{A_n} dx - 2 \int_{\mathbb{R}^N} (u_n^1 + u_n^2) dx + \int_{\mathbb{R}^N} |\nabla (u_n^1 + u_n^2)|^2 dx$$

$$= 2\left(\int_{\mathbb{R}^N} w_{A_n} dx - \int_{\mathbb{R}^N} (u_n^1 + u_n^2) dx \right) + \int_{\mathbb{R}^N} |\nabla(u_n^1 + u_n^2)|^2 dx - \int_{\mathbb{R}^N} |\nabla w_{A_n}|^2 dx.$$

But

$$0 \leq \lim_{n \to \infty} \left| \int_{\mathbb{R}^N} w_{A_n} dx - \int_{\mathbb{R}^N} (u_n^1 + u_n^2) dx \right|$$

$$\leq \lim_{n \to \infty} |A_n|^{\frac{1}{2}} \| w_{A_n} - (u_n^1 + u_n^2) \|_{L^2(A_n)} = 0$$

and

$$\limsup_{n \to \infty} \left[\int_{\mathbb{R}^N} |\nabla(u_n^1 + u_n^2)|^2 dx - \int_{\mathbb{R}^N} |\nabla w_{A_n}|^2 dx \right] \leq 0$$

hence relation (6.21) follows by the Poincaré inequality.

The next lemma contains an important result which establishes a relation between the norm of the difference of two resolvent operators and the norm of the difference of the particular solutions with the right hand side equal to 1.

Lemma 6.4.14 *Let $\tilde{A} \subseteq A$ be two quasi-open sets in $\mathbb{R}^N$ of finite measure. There exist two constants K, α depending only on the measure of A and the dimension of the space, such that*

(6.22) $$\|R_A - R_{\tilde{A}}\|_2 \leq K \|w_A - w_{\tilde{A}}\|_{L^2(\mathbb{R}^N)}^\alpha.$$

Proof   We use in the proof of this lemma some regularity results for $R_A(f)$. We refer to [124] for details concerning the global L^∞-estimation of $R_A(f)$ in terms of f (see also [182]). Although the formulation in [124, Theorem 8.16] is given for open sets, the extension to quasi-open is immediate. For $p > N/2$ we have

(6.23) $$\|R_A(f)\|_{L^\infty} \leq C(p, N, |A|) \|f\|_{L^p}.$$

For $N = 2, 3$, taking $p = 2$ the proof of inequality (6.22) is quite immediate, using the fact that R_A and $R_{\tilde{A}}$ are self-adjoints in $L^2(A)$. For $N > 3$ an interpolation argument has to be used.

Let us fix $p > N/2$, $N \geq 4$. For any $f \in L^p(A)$ with $f \geq 0$, relation (6.23), the maximum principle and the Hölder inequality give

$$\int_A |(R_A - R_{\tilde{A}})(f)|^p dx \leq |(R_A - R_{\tilde{A}})(f)|_{L^\infty}^{p-1} \int_A (R_A - R_{\tilde{A}})(f) dx$$

$$\leq C(p, N, |A|) \|f\|_{L^p}^{p-1} \int_A f(R_A - R_{\tilde{A}})(1) dx$$

$$\leq C(p, N, |A|) \|f\|_{L^p}^{p} \|(R_A - R_{\tilde{A}})(1)\|_{L^{p'}}.$$

Hence

$$\|R_A - R_{\tilde{A}}\|_p \leq 2[C(p, N, |A|)]^{1/p} \|(R_A - R_{\tilde{A}})(1)\|_{L^{p'}}^{1/p}.$$

Denoting by R_A^*, $R_{\tilde{A}}^*$ the adjoint operators of R_A respectively $R_{\tilde{A}}$, we have

$$R_A^* - R_{\tilde{A}}^* : L^{p'}(A) \to L^{p'}(A)$$

and

$$\|R_A^* - R_{\tilde{A}}^*\|_{p'} \leq 2[C(p, N, |A|)]^{1/p} \|(R_A - R_{\tilde{A}})(1)\|_{L^{p'}}^{1/p}.$$

Since both R_A and $R_{\tilde{A}}$ are self-adjoints on $L^2(A)$, keeping the same notations for R_A, $R_{\tilde{A}}$ and their extensions on $L^{p'}(A)$, we get $R_A - R_{\tilde{A}} : L^{p'}(A) \to L^{p'}(A)$ and

$$\|R_A - R_{\tilde{A}}\|_{p'} \leq 2[C(p, N, |A|)]^{1/p} \|(R_A - R_{\tilde{A}})(1)\|_{L^{p'}}^{1/p}.$$

Using the Riesz-Thorin interpolation theorem, we get

$$\|R_A - R_{\tilde{A}}\|_2 \leq 2[C(p, N, |A|)]^{1/p} \|(R_A - R_{\tilde{A}})(1)\|_{L^{p'}}^{1/p}.$$

Since $1 < p' < 2$ we have

$$\|(R_A - R_{\tilde{A}})(1)\|_{L^{p'}}^{1/p} \leq \|(R_A - R_{\tilde{A}})(1)\|_{L^2}^{1/p} |A|^{(2-p')/(pp')}$$

which concludes the proof. ∎

Relation (6.21) and Lemma 6.4.14 yield the following.

Proposition 6.4.15 *If relations (6.17)–(6.20) hold, then*

$$\|R_{A_n} - R_{\tilde{A}_n}\|_2 \to 0 \quad \text{for} \quad n \to \infty.$$

Proof We apply Lemma 6.4.14 to the sequence $\tilde{A}_n \subseteq A_n$. ∎

Proof [of theorem 6.4.8] It is a consequence of Propositions 6.4.11, 6.4.13 and 6.4.15. ∎

Remark 6.4.16 From Theorems 6.4.7 and 6.4.8 it is clear that if the compactness case occurs, for any sequence $v_n \in H_0^1(A_n)$ such that $v_n \rightharpoonup v$ weakly in $H^1(\mathbb{R}^N)$ we have $v_n \longrightarrow v$ strongly $L^2(\mathbb{R}^N)$. If the dichotomy case occurs, we prove that any sequence $v_n \rightharpoonup v$ weakly in $H^1(\mathbb{R}^N)$ "dichotomizes", in the sense that there exist a sequence $\tilde{v}_n \in H_0^1(\tilde{A}_n)$ such that

$$(6.24) \qquad \lim_{n \to \infty} \int_{A_n} (v_n - \tilde{v}_n)^2 dx = 0$$

and

$$(6.25) \qquad \liminf_{n \to \infty} \int_{A_n} [|\nabla v_n|^2 - |\tilde{\nabla} v_n|^2] dx \geq 0.$$

Of course, nothing can be said on the L^2-norm of $\tilde{v}_{n|A_n^1}$ and $\tilde{v}_{n|A_n^2}$.

Indeed, let us prove (6.24) and (6.25) in the dichotomy case. We have that

$$R_A - R_{\tilde{A}} \in \mathcal{L}(L^2(A), H_0^1(A))$$

and by a simple computation

$$(6.26) \qquad \|R_A - R_{\tilde{A}}\|_{\mathcal{L}(L^2(A), H_0^1(A))} \leq 2\sqrt{\|R_A - R_{\tilde{A}}\|_2}.$$

First, for a positive function $f \in L^2(A)$ we have

$$\int_A |\nabla (R_A - R_{\tilde{A}})(f)|^2 dx = \int_A f(R_A - R_{\tilde{A}})(f) dx \leq \|f\|_{L^2(A)}^2 \|R_A - R_{\tilde{A}}\|_2.$$

Taking an arbitrary $f \in L^2(A)$ and decomposing it in $f = f^+ - f^-$ we get (6.26). Hence $(R_A - R_{\tilde{A}})^* \in \mathcal{L}(H^{-1}(A), L^2(A))$, and since $R_A - R_{\tilde{A}}$ is self-adjoint in $L^2(A)$, still denoting $R_A - R_{\tilde{A}}$ its extension on $H^{-1}(A)$ we can write

$$(6.27) \qquad \|R_A - R_{\tilde{A}}\|_{\mathcal{L}(H^{-1}(A), L^2(A))} \leq 2\sqrt{\|R_A - R_{\tilde{A}}\|_2}.$$

Consider an arbitrary sequence $v_n \in H_0^1(A_n)$, such that $v_n \rightharpoonup v$ weakly in $H^1(\mathbb{R}^N)$. Taking $f_n = -\Delta v_n \in H^{-1}(A_n)$ we get

$$\|R_{A_n}(f_n) - R_{\tilde{A}_n}(f_n)\|_{L^2(A_n)} = \|v_n - P_{H_0^1(\tilde{A}_n)} v_n\|_{L^2(A_n)}.$$

Since $\|f_n\|_{H^{-1}(A_n)}$ is bounded, relation (6.27) written for A_n and $\tilde{A}_n$ gives

$$\|v_n - P_{H_0^1(\tilde{A}_n)}v_n\|_{L^2(A_n)} \to 0.$$

Denoting $\tilde{v}_n = P_{H_0^1(\tilde{A}_n)}v_n \in H_0^1(\tilde{A}_n)$ we conclude the proof, since relation (6.25) is an immediate consequence of the fact that the projector's norm is equal to 1.

An important consequence of Proposition 6.4.15 is the following.

Corollary 6.4.17 *If relations (6.17)–(6.20) hold, then for any $k \in \mathbb{N}^*$ we have*

$$\left| \frac{1}{\lambda_k(A_n)} - \frac{1}{\lambda_k(\tilde{A}_n)} \right| \to 0 \quad \text{for } n \to \infty.$$

Proof It follows from Proposition 6.4.15 and [109, Corollary XI.9.4].   ∎

Remark 6.4.18 Let us suppose that a sequence of quasi-open sets of uniformly bounded measure $(A_n)_{n \in \mathbb{N}}$ γ-converges to a measure μ. If injection (6.5) is compact, then we proved that R_{A_n} converges in $\mathcal{L}(L^2(\mathbb{R}^N))$ to R_μ. The converse is also true. Indeed, let us suppose that R_{A_n} converges in $\mathcal{L}(L^2(\mathbb{R}^N))$ to some operator R. Since by hypothesis A_n γ-converges to μ, we immediately have $R = R_\mu$ and the regular set of the measure μ has also finite Lebesgue measure. We have

$$\|R_{A_n}(\max\{\chi_{A_n}, \chi_{A_\mu}\}) - R_{A_\mu}(\max\{\chi_{A_n}, \chi_{A_\mu}\})\|_{L^2(\mathbb{R}^N)}$$

$$\leq \|R_{A_n} - R_{A_\mu}\|_{\mathcal{L}(L^2(\mathbb{R}^N))} \|\max\{\chi_{A_n}, \chi_{A_\mu}\}\|_{L^2(\mathbb{R}^N)}$$

hence it vanishes as $n \to \infty$. This means that w_{A_n} converges strongly in $L^2(\mathbb{R}^N)$, hence Proposition 6.4.11 can be applied.

Optimization of eigenvalues in unbounded design regions. Examples. Let us consider, like in Section 6.3 the problem of minimizing functionals depending on the first and the second eigenvalues in $\mathbb{R}^N$. We define the set

(6.28) $E = \{(\lambda_1(A), \lambda_2(A)) : A \in \mathcal{A}_c(\mathbb{R}^N)\} \subseteq \mathbb{R}^2,$

where $\mathcal{A}_c(\mathbb{R}^N)$ is the family of all quasi-open sets of measure less than or equal to c of $\mathbb{R}^N$. Using the same arguments as in Section 6.3 and the concentration-compactness principle for the γ-convergence, we prove the following theorem.

Theorem 6.4.19 *The set E is closed in $\mathbb{R}^2$.*

Proof Let $(x, y) \in \mathbb{R}^2$ be such that

$$x = \lim_{n \to \infty} \lambda_1(A_n), \quad y = \lim_{n \to \infty} \lambda_2(A_n), \quad \text{with } (A_n)_{n \in \mathbb{N}} \subseteq \mathcal{A}_c(\mathbb{R}^N).$$

In order to prove the closedness of the set E, we have to prove the existence of a set $A \in \mathcal{A}_c(\mathbb{R}^N)$ such that $x = \lambda_1(A), y = \lambda_2(A)$. Using lemma 6.4.8 we distinguish between two situations. We begin with the dichotomy.

There exists a sequence $\tilde{A}_n = A_n^1 \cup A_n^2$ given by Lemma 6.4.8, which in particular satisfies $\lambda_1(A_n^1 \cup A_n^2) \to x$ and $\lambda_2(A_n^1 \cup A_n^2) \to y$. There are two possibilities (up to a renotation of the indices).

1. $\lambda_1(A_n^1) \to x$ and $\lambda_2(A_n^1) \to y$;

2. $\lambda_1(A_n^1) \to x$ and $\lambda_1(A_n^2) \to y$.

If the first situation occurs, there exists $\varepsilon > 0$ such that for every $n \geq n_\varepsilon$, we have $|A_n^1| \leq c - \varepsilon$. For every $\delta > 0$, there exists $n_\delta \in \mathbb{N}$ such that for every $n \geq n_\delta$ we have $|\lambda_1(A_n^1) - x| + |\lambda_2(A_n^1) - y| \leq \delta$. For every $\delta' > 0$, there exists $r > 0$ large enough such that

$$\lambda_1(A_n^1 \cap B_r) - \lambda_1(A_n^1) \leq \delta,$$
$$\lambda_2(A_n^1 \cap B_r) - \lambda_2(A_n^1) \leq \delta,$$
$$|A_n^1 \setminus (A_n^1 \cap B_r)| \leq \delta'.$$

Choosing $\delta > 0$ and $\delta' > 0$ such that

$$\left(\frac{c - \varepsilon + \delta'}{c}\right)^{2/n} < \frac{y}{y + 2\delta}$$

we make an expansion of ratio $\left(\frac{c}{c - \varepsilon + \delta'}\right)^{1/n}$ to $A_n^1 \cap B_r$ for some $n \geq \max\{n_\varepsilon, n_\delta\}$ and find a bounded quasi-open set $A^* = \left(\frac{c}{c - \varepsilon + \delta'}\right)^{1/n} (A_n^1 \cap B_r)$ such that $|A^*| \leq c$ and $\lambda_1(A^*) \leq x, \lambda_2(A^*) \leq y$.

If the second situation occurs, we replace A_n^1 by the ball of mass $|A_n^1|$ denoted B_n^1 and A_n^2 by the ball of mass $|A_n^2|$ denoted B_n^2. For a subsequence (still denoted with the same indices) we find two balls B_1, B_2, such that $|B_1| + |B_2| \leq c$, and γ-limits of $(B_n^1)_{n \in \mathbb{N}}, (B_n^2)_{n \in \mathbb{N}}$ are respectively B_1 and B_2.

Then $\lambda_1(B_1 \cup B_2) \leq x, \lambda_2(B_1 \cup B_2) \leq y$ and the same arguments as in the bounded case can be applied.

If compactness occurs when applying Lemma 6.4.8, there are two possibilities: either A_μ is quasi-connected, or not. If A_μ is not quasi-connected, then we write $A_\mu = A_1 \cup A_2$ and repeat the same arguments as in the dichotomy situation. If A_μ is quasi-connected, two possibilities may occur.

1. $\lambda_1(A_\mu) = \lambda_1(\mu)$,

2. $\lambda_1(A_\mu) < \lambda_1(\mu)$

If the first situation occurs, we get $A_\mu = \mu$ (see [42, Proposition 3.3]). Indeed, let us denote by u_μ a first eigenvector associated to $\lambda_1(\mu)$. Since $u_\mu \in H_0^1(A_\mu)$ we get

$$\lambda_1(\mu) = \frac{\int_{\mathbb{R}^N} |\nabla u_\mu|^2 dx + \int_{\mathbb{R}^N} u_\mu^2 d\mu}{\int_{\mathbb{R}^N} u_\mu^2 dx} = \frac{\int_{\mathbb{R}^N} |\nabla u_\mu|^2 dx}{\int_{\mathbb{R}^N} u_\mu^2 dx} = \lambda_1(A_\mu).$$

Consequently, $\int_{\mathbb{R}^N} u_\mu^2 d\mu = 0$ and u_μ is a first eigenvector for A_μ. Since A_μ is quasi-connected, we get that $u_\mu(x) > 0$ q.e. $x \in A_\mu$. Then from $\int_{\mathbb{R}^N} u_\mu^2 d\mu = 0$ we get $\mu(A_\mu) = 0$, hence $\mu = A_\mu$. Thus, $\lambda_1(A_\mu) = x, \lambda_2(A_\mu) = y$.

If the second situation occurs, there are two possibilities: either $\lambda_2(A_\mu) < y$, or $\lambda_2(A_\mu) = y$. If $\lambda_2(A_\mu) < y$ we can consider a ball B_r large enough such that $\lambda_1(A_\mu \cap B_r) < x, \lambda_2(A_\mu \cap B_r) < y$ and follow the same arguments as in the bounded case.

If $\lambda_2(A_\mu) = y$, let us denote by u_2 an eigenvector associated to the second eigenvalue, and $\tilde{A}_\mu = \{u \neq 0\}$. It follows like in the bounded case [38] that

$$\lambda_1(\tilde{A}_\mu) = \lambda_2(\tilde{A}_\mu) = y.$$

There exists a mapping, like in the bounded case (see Section 6.3)

$$[0, +\infty] \ni t \mapsto A(t) \in \mathcal{A}_c(\mathbb{R}^N)$$

with the properties $A(0) = A_\mu$, $A(+\infty) = \tilde{A}_\mu$, for every $t_1 < t_2$, $\text{Cap}(A_{t_2} \setminus A_{t_1}) = 0$, the mapping $t \mapsto \lambda_1(A(t))$ is continuous and increasing and the mapping $t \mapsto \lambda_2(A(t))$ is constant. The idea to prove this assertion is to "delete" continuously in capacity the nodal line of u_2. We give not the proof here, since the passage from bounded to unbounded set can be done by classical arguments.

Then, there exists some $t \in (0, +\infty)$ such that $\lambda_1(A_t) = x, \lambda_2(A_t) = y$ and $A \in \mathcal{A}_c(\mathbb{R}^N)$. ∎

6.5 Some open questions

- Concerning Theorem 6.3.2, there are many other questions which can be raised. Is the set E convex? Is E still closed if the pair (λ_1, λ_2) is replaced by (λ_i, λ_j), or more generally if we consider the set

$$E_K = \left\{ (\lambda_i(A))_{i \in K} : \quad A \in \mathcal{A}_c(B) \right\}$$

where K is a given subset of positive integers? Are the sets A on the boundary of E smooth? When they are convex? If the design region is an open set D, is the set $s(\mathcal{A}_c(D))$ still closed? Or if the Laplace operator is replaced by

$$L = -\partial_i(a_{ij}\partial_j) + b_i\partial_i + c \quad ?$$

- If instead of a shape functional depending only on λ_1, λ_2 one considers the first k-th eigenvalues $\lambda_1, \lambda_2, ..., \lambda_k$, is the analogous version of Theorem 6.3.2 true?

- In [43] it was proved that problem

(6.29) $$\min\{\lambda_k(\Omega) : \Omega \subseteq \mathbb{R}^N, |\Omega| = c\}$$

has a solution for $k = 3$. In two dimensions, the disk is suspected to be the solution (see [14]); up to now, as far as we know, this is still a conjecture. The existence for (6.29) in the case $k \geq 4$ is not solved. Roughly speaking, if one proves the existence of bounded minimizers for $\lambda_3, ..., \lambda_k$ (under the volume constraint) then the existence of a minimizer (bounded or unbounded) for λ_{k+1} follows.

- An important question concerns the regularity of the optimal shape. Is there a smooth solution for Problem (6.29) (even for $k = 3$)? Only for the first eigenvalue is known that, for constant volume in a bounded design region, the minimizer is open and not quasi-open (see [132]).

For a detailed list of open problems related to eigenvalues we refer the reader to [14].

Bibliography

[1] E. ACERBI, G. BUTTAZZO: *Reinforcement problems in the calculus of variations. (English. Ann. Inst. H. Poincar Anal. Non Linaire 3 (1986), no. 4, 273–284.

[2] R. A. ADAMS: *Sobolev Spaces.* Academic Press, New York (1975).

[3] D.R. ADAMS, L.I. HEDBERG: *Function Spaces and Potential Theory.* Springer-Verlag Berlin Heidelberg (1996).

[4] G. ALLAIRE: *Shape optimization by the homogeneization method,* Applied Mathematical Sciences, 146. Springer-Verlag, New York (2002).

[5] G. ALLAIRE, E. BONNETIER, G. FRANCFORT, F.JOUVE: *Shape optimization by the homogenization method.* Numer. Math. **76** (1997), no. 1, 27–68.

[6] G. ALLAIRE, R. V. KOHN: *Optimal design for minimum weight and compliance in plane stress using extremal microstructures.* Europ. J. Mech. A/Solids, **12** (6) (1993), 839–878.

[7] L. AMBROSIO: *Existence theory for a new class of variational problems.* Arch. Ration. Mech. Anal., **111** (4) (1990), 291–322.

[8] L. AMBROSIO: *Introduction to geometric measure theory and minimal surfaces* (Italian), Scuola Norm. Sup., Pisa (1997).

[9] L. AMBROSIO, A. BRAIDES: *Functionals defined on partitions of sets of finite perimeter, I: integral representation and Γ-convergence.* J. Math. Pures Appl., **69** (1990), 285–305.

[10] L. AMBROSIO, A. BRAIDES: *Functionals defined on partitions of sets of finite perimeter, II: semicontinuity, relaxation and homogenization.* J. Math. Pures Appl., **69** (1990), 307–333.

[11] L. AMBROSIO, G. BUTTAZZO: *An optimal design problem with perimeter penalization.* Calc. Var., **1** (1993), 55–69.

[12] L. AMBROSIO, N. FUSCO, D. PALLARA: *Functions of bounded variation and free discontinuity problems.* Oxford Mathematical Monographs, Clarendon Press, Oxford (2000).

[13] M. S. ASHBAUGH, R. D. BENGURIA: *Proof of the Payne-Pólya-Weinberger conjecture.* Bull. Amer. Math. Soc., **25** (1991), 19–29.

[14] M. S. ASHBAUGH: *Open problems on eigenvalues of the Laplacian.* Analytic and geometric inequalities and applications, 13–28, Math. Appl., 478, Kluwer Acad. Publ., Dordrecht (1999).

[15] H. ATTOUCH: *Variational Convergence for Functions and Operators.* Pitman, Boston, 1984.

[16] H. ATTOUCH, G. BUTTAZZO: *Homogenization of reinforced periodic one-codimensional structures.* Ann. Scuola Norm. Sup. Pisa Cl. Sci., **14** (1987), 465–484.

[17] H. ATTOUCH, C. PICARD: *Variational inequalities with varying obstacles: The general form of the limit problem.* J. Funct. Anal., **50** (1983), 329–386.

[18] D. AZE, G. BUTTAZZO: *Some remarks on the optimal design of periodically reinforced structures.* RAIRO Modél. Math. Anal. Numér., **23** (1989), 53–61.

[19] J. BAXTER, G. DAL MASO: *Stopping times and Γ-convergence,* Trans. American Math. Soc. **vol. 303**, 1, September (1987), 1–38.

[20] M. BELLONI, G. BUTTAZZO, L. FREDDI: *Completion by Gamma-convergence for optimal control problems.* Ann. Fac. Sci. Toulouse Math., **2** (1993), 149–162.

[21] M. BELLONI, B. KAWOHL: *A paper of Legendre revisited.* Forum Math. 9 (1997), no. 5, 655–667.

[22] M. BELLONI, A. WAGNER: *Newton's problem of minimal resistance in the class of bodies with prescribed volume.* Preprint 2001.

[23] M. BENDSØE: *Optimal shape design as a material distribution problem.* Struct. Optim., **1** (1989), 193–202.

[24] G. BOUCHITTE, G. BUTTAZZO: *Characterization of optimal shapes and masses through Monge-Kantorovich equation.* Preprint Dipartimento di Matematica Università di Pisa, Pisa (2000).

[25] G. BOUCHITTE, G. BUTTAZZO, I. FRAGALÀ: *Convergence of Sobolev spaces on varying manifolds.* Preprint Dipartimento di Matematica Università di Pisa, Pisa (1999).

[26] G. BOUCHITTE, G. BUTTAZZO, P. SEPPECHER: *Energies with respect to a measure and applications to low dimensional structures.* Calc. Var., **5** (1997), 37–54.

[27] G. BOUCHITTE, G. BUTTAZZO, P. SEPPECHER: *Shape optimization solutions via Monge-Kantorovich equation.* C. R. Acad. Sci. Paris, **324-I** (1997), 1185–1191.

[28] G. BOUCHITTE, G. DAL MASO: *Integral representation and relaxation of convex local functionals on $BV(\Omega)$.* Ann. Scuola Norm. Sup. Pisa Cl. Sci., (4) **20** (1993), 483–533.

[29] H.J. BRASCAMP, E. LIEB, J.M. LUTTINGER: *A general rearrangement inequality for mulptiple integrals,* J. Funct. Anal., **17** (1974), 227–237.

[30] H. BREZIS: *Analyse Fonctionelle.* Masson, Paris (1983).

[31] H. BREZIS, L. CAFFARELLI, A. FRIEDMAN: *Reinforcement problems for elliptic equations and variational inequalities,* Ann. Mat. Pura Appl. (4) **123** (1980), 219–246.

[32] F. BROCK: *Continuous Steiner Symmetrization.* Math. Nachrichten, **172** (1995), 25–48.

[33] F. BROCK, V. FERONE, B. KAWOHL: *A symmetry problem in the calculus of variations.* Calc. Var. Partial Differential Equations 4 (1996), no. 6, 593–599.

[34] D. BUCUR: *Shape analysis for Nonsmooth Elliptic Operators.* Appl. Math. Lett., **9** (3) (1996), 11–16.

[35] D. BUCUR: *Characterization for the Kuratowski Limits of a Sequence of Sobolev Spaces.*, J. Differential Equations, **151** (1999), 1–19.

[36] D. BUCUR: *Uniform concentration-compactness for Sobolev spaces on variable domains,* J. Differential Equations, **162 (2)** (2000), 427–450.

[37] D. BUCUR, G. BUTTAZZO: *Results and questions on minimum problems for eigenvalues.* Preprint Dipartimento di Matematica Università di Pisa, Pisa (1998).

[38] D. BUCUR, G. BUTTAZZO, I. FIGUEIREDO: *On the attainable eigenvalues of the Laplace operator.* SIAM J. Math. Anal., **30** (1999), 527–536.

[39] D. BUCUR, G. BUTTAZZO, A. HENROT: *Existence results for some optimal partition problems.* Adv. Math. Sci. Appl., **8** (1998), 571–579.

[40] D. BUCUR, G. BUTTAZZO, P. TREBESCHI: *An existence result for optimal obstacles.* J. Funct. Anal., **162** (1999), 96–119.

[41] D. BUCUR, G. BUTTAZZO, N. VARCHON: *On the problem of optimal cutting,* SIAM J. Optimization (to appear).

[42] D. BUCUR, T. CHATELAIN: *Strict monotonicity of the second eigenvalue of the Laplace operator on relaxed domains,* Bull. for Appl. and Comp. Math. **1510–1566** (1998), 115–122.

[43] D. BUCUR, A. HENROT: *Minimization of the third eigenvlaue of the Dirichlet Laplacian.* Proc. Roy. Soc. London, Ser. A, **456** (2000), 985–996.

[44] D. BUCUR, A. HENROT: *Stability for the Dirichlet problem under continuous Steiner symmetrization.* Potential Analysis. **13** (2000), 127–145.

[45] D. BUCUR, A. HENROT, J. SOKOLOWSKI, A. ZOCHOWSKI: *Continuity of the elasticity system solutions with respect to boundary variations.* Adv. Math. Sci. Appl. **11** (2001), no. 1, 57–73.

[46] D. BUCUR, P. TREBESCHI: *Shape optimization problem governed by nonlinear state equation.* Proc. Roy. Soc. Edinburgh, **128 A** (1998), 945–963.

[47] D. BUCUR, N. VARCHON: *Boundary variation for the Neumann problem ,* Ann. Scuola Norm. Sup. Pisa Cl. Sci.**XXIV (4)** (2000), 807–821.

[48] D. BUCUR, N. VARCHON: *A duality approach for the boundary variations of Neumann problems,* Preprint Université de Franche-Comté n. 00/14, Becançon (2000).

[49] D. BUCUR, J. P. ZOLESIO *Wiener's criterion and shape continuity for the Dirichlet problem.* Boll. Un. Mat. Ital. B (7) 11 (1997), no. 4, 757–771.

[50] D. BUCUR, J. P. ZOLESIO: *N-Dimensional Shape Optimization under Capacitary Constraints.* J. Differential Equations, **123** (2) (1995), 504–522.

[51] D. BUCUR, J. P. ZOLESIO: *Shape continuity for Dirichlet-Neumann problems.* Progress in partial differential equations: the Metz surveys, 4, 53–65, Pitman Res. Notes Math. Ser., 345, Longman, Harlow (1996).

[52] H. BUSEMAN, G. EWALD, G. C. SHEPARD: *Convex bodies and convexity in Grassman cones, I-IV.* Math.Ann., **151**, (1963), 1–41.

[53] G. BUTTAZZO: *Thin insulating layers: the optimization point of view.* Proceedings of "Material Instabilities in Continuum Mechanics and Related Mathematical Problems", Edinburgh 1985–1986, edited by J. M. Ball, Oxford University Press, Oxford (1988), 11–19.

[54] G. BUTTAZZO: *Semicontinuity, Relaxation and Integral Representation in the Calculus of Variations.* Pitman Res. Notes Math. Ser. **207**, Longman, Harlow (1989).

[55] G. BUTTAZZO: *Relaxed formulation for a class of shape optimization problems.* Proceedings of "Boundary Control and Boundary Variations", Sophia-Antipolis, October 15–17, 1990, Lecture Notes in Control and Inf. Sci. **178**, Springer-Verlag, Berlin (1992), 50–59.

[56] G. BUTTAZZO: *Existence via relaxation for some domain optimization problems.* Proceedings of "Topology Design of Structures", Sesimbra 20–26 June 1992, NATO ASI Series E: Applied Sciences **227**, Kluwer Academic Publishers, Dordrecht (1993), 337–343.

[57] G. BUTTAZZO: *Γ-convergence and its applications to some problems in the calculus of variations.* Lecture notes of a series of lectures held in "School of Homogenization", ICTP, Trieste 6–17 September 1993. Printed by SISSA, Trieste (1993), 81–106.

[58] G. BUTTAZZO: *Relaxed shape optimization problems with Dirichlet boundary conditions.* Proceedings of "Composite Media and Homogenization Theory II", Trieste 20 September – 1 October 1993, World Scientific, Singapore (1995), 125–137.

[59] G. BUTTAZZO: *Semicontinuità inferiore di funzionali definiti su BV.* Lecture notes of a series of lectures held in "Scuola Internazionale di Equazioni Differenziali e Calcolo delle Variazioni", Pisa 14–26 September 1992, Pitagora Editrice, Bologna (1995), 197–235.

[60] G. BUTTAZZO: *A general theory of relaxed controls and application to shape optimization problems*. Proceedings of "Progress in Elliptic and Parabolic Partial Differential Equations", Capri September 19–23 1994, Pitman Res. Notes Math. Ser. **350**, Longman, Harlow (1996), 58–70.

[61] G. BUTTAZZO: *On the existence of minimizing domains for some shape optimization problems*. Proceedings of "29ème Congrès d'Analyse Numérique - CANum'97", Domaine d'Imbours May 26–30 1997, ESAIM Proceedings, **3** (1998), 53–65.

[62] G. BUTTAZZO, G. DAL MASO: *Shape optimization for Dirichlet problems: relaxed solutions and optimality conditions*. Bull. Amer. Math. Soc., **23** (1990), 531–535.

[63] G. BUTTAZZO, G. DAL MASO: *Shape optimization for Dirichlet problems: relaxed formulation and optimality conditions*. Appl. Math. Optim., **23** (1991), 17–49.

[64] G. BUTTAZZO, G. DAL MASO: *An existence result for a class of shape optimization problems*. Arch. Rational Mech. Anal., **122** (1993), 183–195.

[65] G. BUTTAZZO, G. DAL MASO, A. GARRONI, A. MALUSA: *On the relaxed formulation of Some Shape Optimization Problems*. Adv. Math. Sci. Appl., **7** (1997), 1–24.

[66] G. BUTTAZZO, V. FERONE, B. KAWOHL: *Minimum problems over sets of concave functions and related questions*. Math. Nachr., **173** (1995), 71–89.

[67] G. BUTTAZZO, L. FREDDI: *Relaxed optimal control problems and applications to shape optimization*. Lecture notes of a course held at the NATO-ASI Summer School "Nonlinear Analysis, Differential Equations and Control", Montreal, July 27 – August 7, 1998, Kluwer, Dordrecht (1999), 159–206.

[68] G. BUTTAZZO, M. GIAQUINTA, S. HILDEBRANDT: *One-dimensional Calculus of Variations: an Introduction*. Oxford University Press, Oxford (1998).

[69] G. BUTTAZZO, P. GUASONI: *Shape optimization problems over classes of convex domains*. J. Convex Anal., **4** (1997), 343–351.

[70] G. BUTTAZZO, B. KAWOHL: *On Newton's problem of minimal resistance*. Math. Intelligencer, **15** (1993), 7–12.

[71] G. BUTTAZZO, P. TREBESCHI: *The role of monotonicity in some shape optimization problems.* In "Calculus of Variations, Differential Equations and Optimal Control", Research Notes in Mathematics Series, Vol. 410-411, Chapman & Hall/CRC Press, Boca Raton, (1999).

[72] G. BUTTAZZO, A. WAGNER: *On the optimal shape of a rigid body supported by an elastic membrane.* Nonlinear Anal., **39** (2000), 47–63.

[73] G. BUTTAZZO, O. M. ZEINE: *Un problème d'optimisation de plaques.* Modél. Math. Anal. Numér., **31** (1997), 167–184.

[74] E. CABIB: *A relaxed control problem for two-phase conductors.* Ann. Univ. Ferrara - Sez. VII - Sc. Mat., **33** (1987), 207–218.

[75] E. CABIB, G. DAL MASO: *On a class of optimum problems in structural design.* J. Optimization Theory Appl., **56** (1988), 39–65.

[76] G. CARLIER, T. LACHAND-ROBERT: sl Regularity of solutions for some variational problems subject to a convexity constraint. Comm. Pure Appl. Math. 54 (2001), no. 5, 583–594.

[77] J. CEA, K. MALANOWSKI: *An example of a max-min problem in partial differential equations.* SIAM J. Control, **8** (1970), p. 305–316.

[78] A. CHAMBOLLE, F. DOVERI: *Continuity of Neumann linear elliptic problems on varying two-dimensional bounded open sets.*, Commun. Partial Differ. Equations, **22** (5-6) (1997), 811–840.

[79] A. CHAMBOLLE: *A density result in two-dimensional linearized elasticity and applications,* Preprint Ceremade 2001.

[80] D. CHENAIS: *On the existence of a solution in a domain identification problem.* J. Math. Anal. Appl., **52** (1975), 189–219.

[81] D. CHENAIS: *Homéomorphisme entre ouverts lipschitziens.* Ann. Mat. Pura Appl. (4) 118 (1978), 343–398.

[82] M. CHIPOT, G. DAL MASO: *Relaxed shape optimization: the case of nonnegative data for the Dirichlet problem.* Adv. Math. Sci. Appl., **1** (1992), 47–81.

[83] D. CIORANESCU, F. MURAT: *Un terme trange venu d'ailleurs.* Nonlinear partial differential equations and their applications. Collge de France Seminar, Vol. II (Paris, 1979/1980), pp. 98–138, 389–390, Res. Notes in Math., 60, Pitman, Boston, Mass.-London (1982).

[84] S. COX, B. KAWOHL: *Circular symmetrization and extremal Robin conditions*, Z. Angew. Math. Phys., **50** (1999), 301–311 .

[85] M. COMTE, T. LACHAND-ROBERT: *Existence of minimizers for Newton's problem of the body of minimal resistance under a single impact assumption.* J. Anal. Math. **83** (2001), 313–335.

[86] M. COMTE, T. LACHAND-ROBERT: *Newton's problem of the body of minimal resistance under a single-impact assumption.* Calc. Var. Partial Differential Equations **12** (2001), no. 2, 173–211.

[87] R. COURANT, D. HILBERT: *Methods of mathematical physics.* Vol. I. Interscience Publishers, Inc., New York, N.Y. (1953).

[88] B. DACOROGNA: *Direct Methods in the Calculus of Variations.* Appl. Math. Sciences **78**, Springer-Verlag, Berlin (1989).

[89] G. DAL MASO: *An Introduction to Γ-convergence.* Birkhäuser, Boston (1993).

[90] G. DAL MASO: *On the integral representation of certain local functionals.* Ricerche Mat., **32** (1983), 85–113.

[91] G. DAL MASO: *Some necessary and sufficient conditions for convergence of sequences of unilateral convex sets.* J. Funct. Anal, **62** (1985), 119–159.

[92] G. DAL MASO: *Γ-convergence and μ-capacities.* Ann. Scuola Norm. Sup. Pisa, **14** (1988), 423–464.

[93] G. DAL MASO, A. DE FRANCESCHI: *Limits of nonlinear Dirichlet problems in varying domains.* Manuscripta Math., **61** (1988), 251–268.

[94] G. DAL MASO, A. GARRONI: *New results on the asymptotic behaviour of Dirichlet problems in perforated domains.* Math. Mod. Meth. Appl. Sci., **3** (1994), 373–407.

[95] G. DAL MASO, P. LONGO: *Γ-limits of obstacles.* Ann. Mat. Pura Appl., **128** (1980), 1–50.

[96] G. DAL MASO, A. MALUSA: *Approximation of relaxed Dirichlet problems by boundary value problems in perforated domains.* Proc. Roy. Soc. Edinburgh Sect. **A 125** (1995), no. 1, 99–114.

[97] G. DAL MASO, U. MOSCO: *Wiener's criterion and Γ-convergence.* Appl. Math. Optim., **15** (1987), 15–63.

[98] G. DAL MASO, U. MOSCO: *The Wiener modulus of a radial measure.* Houston J. Math., **15** (1989), 35–57.

[99] G. DAL MASO, U. MOSCO: *Wiener criteria and energy decay for relaxed Dirichlet problems.* Arch. Rational Mech. Anal., **95** (1986), 345–387.

[100] G. DAL MASO, F. MURAT: *Asymptotic behavior and correctors for Dirichlet problems in perforated domains with homogeneous monotone operators.* Ann. Scuola Norm. Sup. Pisa, **24** (1997), 239–290.

[101] G. DAL MASO, R. TOADER: *A model for the quasi-static growth of a brittle fracture: existence and approximation results,* Preprint SISSA 2001.

[102] E. DE GIORGI: *Teoremi di semicontinuità nel calcolo delle variazioni.* Notes of a course given at the Instituto Nazionale de Alta Matematica, Rome (1968).

[103] E. DE GIORGI: *Γ-convergenza e G-convergenza.* Boll. Un. Mat. Ital., (5) **14-A** (1977), 213–224.

[104] E. DE GIORGI, F. COLOMBINI, L.C. PICCININI: *Frontiere orientate di misura minima e questioni collegate.* Quaderni della Scuola Normale Superiore, Pisa, (1972).

[105] E. DE GIORGI, G. DAL MASO, P. LONGO: *Γ-limiti di ostacoli.* Atti Accad. Naz. Lincei Rend. Cl. Sci. Fis. Mat. Natur., (8) **68** (1980), 481–487.

[106] E. DE GIORGI, T. FRANZONI: *Su un tipo di convergenza variazionale.* Atti Accad. Naz. Lincei Cl. Sci. Fis. Mat. Natur., (8) **58** (1975), 842–850.

[107] M. DELFOUR, J.-P. ZOLESIO: *Shapes and geometries. Analysis, differential calculus, and optimization,* Advances in Design and Control (SIAM), Philadelphia, PA (2001).

[108] F. DEMENGEL, R. TEMAM: *Convex functions of measures and applications.* Indiana Univ. Math. J., **33** (1984), 673–709.

[109] N. DUNFORD, J.T. SCHWARTZ: *Linear Operators, Part II: Spectral Theory,* Interscience Publishers, New York, London, (1963).

[110] I. EKELAND, R. TEMAM: *Convex Analysis and Variational Problems.* Studies in Mathematics and its Applications **1**, North-Holland, Amsterdam (1976).

[111] M.J. ESTEBAN, P.L. LIONS: Γ-convergence and the concentration-compactness method for some variational problems with lack of compactness. Ricerche di Matematica, **XXXVI**, fasc. 1, (1987), 73–101.

[112] L. C. EVANS: Partial differential equations and Monge-Kantorovich mass transfer. Current Developments in Mathematics, Cambridge MA (1997), 65–126, Int. Press, Boston (1999).

[113] L. C. EVANS, W. GANGBO: Differential Equations Methods for the Monge-Kantorovich Mass Transfer Problem. Mem. Amer. Math. Soc. **137**, Providence (1999).

[114] L. C. EVANS, R. F. GARIEPY: Measure Theory and Fine Properties of Functions. Studies in Advanced Math., CRC Press, Ann Harbor (1992).

[115] G. FABER: Beweis, dass unter allen homogenen Membranen von gleicher Fläche und gleicher Spannung die kreisförmige den tiefsten Grundton gibt. Sitz. Ber. Bayer. Akad. Wiss., (1923),169–172.

[116] H. FEDERER: Geometric Measure Theory. Springer-Verlag, Berlin (1969).

[117] S. FINZI VITA: Numerical shape optimization for relaxed Dirichlet problems. Preprint Università di Roma "La Sapienza", Roma (1990).

[118] S. FINZI VITA: Constrained shape optimization for Dirichlet problems: discretization via relaxation. Preprint Universita di Roma, **42** (1996).

[119] G. A. FRANCFORT, F. MURAT: Homogenization and optimal bounds in linear elasticity. Arch. Rational Mech. Anal., **94** (1986), 307-334.

[120] J. FREHSE: Capacity methods in the theory of partial differential equations. Jahresber. Deutsch. Math.-Verein. 84 (1982), no. 1, 1-44.

[121] B. FUGLEDE: Finely harmonic functions, Lecture notes in Math., **289**, Springer, Berlin-Heidelberg-New York (1972).

[122] W. GANGBO, R. J. McCANN: The geometry of optimal transportation. Acta Math., **177** (1996), 113–161.

[123] S. GARREAU, P. GUILLAUME, M. MASMOUDI: The topological asymptotic for PDE systems: the elasticity case. SIAM J. Control Optim. **39** (2001), no. 6, 1756–1778.

[124] D. GILBARG, N.S. TRUDINGER: *Elliptic partial differential equations of second order*, Springer-Verlag,Berlin, Heidelberg, New York, Tokyo, (1983).

[125] V. GIRAULT, P.A. RAVIART: *Finite element methods for Navier-Stokes Equations*, Springer-Verlag, Berlin 1986.

[126] E. GIUSTI: *Minimal Surfaces and Functions of Bounded Variation*. Birkäuser, Boston (1984).

[127] E. GIUSTI: *Metodi diretti nel calcolo delle variazioni*. Unione Matematica Italiana, Bologna (1994).

[128] C. GOFFMAN, J. SERRIN: *Sublinear functions of measures and variational integrals*. Duke Math. J., **31** (1964), 159–178.

[129] F. GOLAY, P. SEPPECHER: *Locking materials and topology of optimal shapes*. Paper in preparation. ???

[130] H. H. GOLDSTINE: *A History of the Calculus of Variations from the 17th through the 19th Century*. Springer-Verlag, Heidelberg (1980).

[131] P. GUASONI: *Problemi di ottimizzazione di forma su classi di insiemi convessi*. Tesi di Laurea, Università di Pisa, 1995-1996.

[132] M. HAYOUNI: *Sur la minimisation de la premire valeur propre du laplacien*. C. R. Acad. Sci. Paris Sr. I Math. 330 (2000), no. 7, 551–556.

[133] L.I. HEDBERG: *Spectral Synthesis in Sobolev Spaces and Uniqueness of Solutions of Dirichlet Problems*. Acta Math., **147** (1982), 237–263.

[134] A. HENROT: *Continuity with respect to the domain for the Laplacian: a survey*. Control and Cybernetics, **23** No.3 (1994), 427–443.

[135] A. HENROT, M. PIERRE: *Optimisation de forme* (book in preparation).

[136] J. HEINONEN, T. KILPELAINEN, O. MARTIO: *Nonlinear potential theory of degenerate elliptic equations*. Clarendon Press, Oxford (1993).

[137] D. HORSTMANN, B. KAWOHL, P. VILLAGGIO: *Newton's aerodynamic problem in the presence of friction*. Preprint 2000.

[138] A. D. IOFFE: *On lower semicontinuity of integral functionals I. II.* SIAM J. Control Optim., **15** (1977), 521–538 and 991–1000.

[139] O. KAVIAN: *Introduction à la théorie des points critiques et applications aux problèmes elliptiques.* Mathématiques et applications, **13**, Springer-Verlag, (1993).

[140] T. KILPELAINEN, J. MALY: *Supersolutions to degenerate elliptic equations on quasi open sets,* Commun. in Partial Diff. Eq., **17** (1992), 371–405.

[141] M.V. KELDYSH: *On the Solvability and Stability of the Dirichlet Problem.* Amer. Math. Soc. Translations, **51-2** (1966), 1–73.

[142] D. KINDERLEHRER, G. STAMPACCHIA: *An Introduction to Variational Inequalities and their Applications.* Academic Press, New York (1980).

[143] R. V. KOHN, G. STRANG: *Optimal design and relaxation of variational problems, I,II,III.* Comm. Pure Appl. Math., **39** (1986), 113–137, 139–182, 353–377.

[144] R. V. KOHN, M. VOGELIUS: *Relaxation of a variational method for impedance computed tomography.* Comm. Pure Appl. Math., **40** (1987), 745–777.

[145] E. KRAHN: *Über eine von Rayleigh formulierte Minimaleigenschaft des Kreises.* Math. Ann., **94** (1924), 97–100.

[146] E. KRAHN: *Über Minimaleigenschaften der Kugel in drei un mehr Dimensionen.* Acta Comm. Univ. Dorpat., **A9** (1926), 1–44.

[147] T. LACHAND-ROBERT, M.A. PELETIER: *An example of non-convex minimization and an application to Newton's problem of the body of least resistance.* Ann. Inst. H. Poincar Anal. Non Linaire **18** (2001), no. 2, 179–198

[148] T. LACHAND-ROBERT, M.A. PELETIER: *Newton's problem of the body of minimal resistance in the class of convex developable functions.* Math. Nachr. 226 (2001), 153–176.

[149] E.H. LIEB: *On the lowest eigenvalue of the Laplacian for the intersection of two domains.* Inventiones matematicae, **74**, (1983), 441–448.

[150] P.L. LIONS: *The concentration-compactness principle in the Calculus of Variations. The locally compact case, part 1.* Ann. Inst. Poincaré, **1**, no. 2 (1984), 109–145.

[151] W. LIU, P. NEITTAANMAKI, D. TIBA: *Sur les problmes d'optimisation structurelle.* C. R. Acad. Sci. Paris Sr. I Math. **331** (2000), no. 1, 101–106.

[152] K. A. LURIE, A. V. CHERKAEV: *G-closure of a Set of Anisotropically Conductivity Media in the Two-Dimensional Case.* J. Optimization Theory Appl., **42** (1984), 283–304.

[153] K. A. LURIE, A. V. CHERKAEV: *G-closure of Some Particular Sets of Admissible Material Characteristics for the Problem of Bending of Thin Elastic Plates.* J. Optimization Theory Appl., **42** (1984), 305–316.

[154] P. MARCELLINI: *Nonconvex integrals of the calculus of variations.* Methods of nonconvex analysis (Varenna, 1989), 16–57, Lecture Notes in Math., 1446, Springer, Berlin, (1990).

[155] U. MASSARI, M. MIRANDA: *Minimal surfaces of codimension one*, North-Holland, Amsterdam (1984).

[156] V. G. MAZ'JA: *Sobolev Spaces.* Springer-Verlag, Berlin (1985).

[157] A. MIELE: *Theory of Optimum Aerodynamic Shapes.* Academic Press, New York (1965).

[158] F. MORGAN: *Geometric Measure Theory, a Beginners Guide.* Academic Press, New York (1988).

[159] C. B. MORREY: *Multiple integrals in the calculus of variations.* Springer, Berlin, (1966).

[160] U. MOSCO: *Convergence of convex sets and of solutions of variational inequalities.* Adv. in Math., **3** (1969), 510–585.

[161] U. MOSCO: *Composite media and asymptotic Dirichlet forms.* J. Funct. Anal., **123** (1994), 368–421.

[162] F. MURAT, J. SIMON: *Sur le contrôle par un domaine géométrique.* Preprint 76015, Univ. Paris VI, (1976).

[163] F. MURAT, L. TARTAR: *Calcul des variations et homogénéisation.* Proceedings of "Les Méthodes de l'homogénéisation: Théorie et applications en physique", Ecole d'Eté d'Analyse Numérique C.E.A.-E.D.F.-INRIA, Bréau-sans-Nappe 1983, Collection de la direction des études et recherches d'electricité de France **57**, Eyrolles, Paris, (1985), 319–369.

[164] F. MURAT, L. TARTAR: *Optimality conditions and homogenization.* Proceedings of "Nonlinear variational problems", Isola d'Elba 1983, Res. Notes in Math. **127**, Pitman, London, (1985), 1–8.

[165] P. NEITTAANMAKI, D. TIBA: *Shape optimization in free boundary systems. Free boundary problems: theory and applications*, II (Chiba, 1999), 334–343, GAKUTO Internat. Ser. Math. Sci. Appl., **14**, Gakkōtosho, Tokyo (2000).

[166] R. OSSERMAN: *Bonnesen-style isoperimetric inequalities.* Amer. Math. Monthly, **86** (1979), 1–29.

[167] O. PIRONNEAU: *Optimal Shape Design for Elliptic Systems.* Springer-Verlag, Berlin, 1984.

[168] G. POLYA: *On the characteristic frequencies of a symmetric membrane,* Math. Zeit., **63** (1955), 331–337.

[169] S. T. RACHEV, L. RÜSCHENDORF: *Mass transportation problems. Vol. I Theory, Vol. II Applications.* Probability and its Applications, Springer-Verlag, Berlin (1998).

[170] Y.G. RESHETNYAK: *Weak convergence of completely additive vector measures on a set.* Sibirskii Math. Zh., **9** (1968), 1039–1045, 1386–1394.

[171] J. RAUCH, M. TAYLOR: *Potential and Scattering Theory on Wildly Perturbed Domains.* J. Funct. Analysis, **18** (1975), 27–59.

[172] R. T. ROCKAFELLAR: *Convex Analysis.* Princeton University Press, Princeton (1972).

[173] C. A. ROGERS: *Hausdorff Measures.* Cambridge University Press, Princeton (1972).

[174] F. SIMONDON: *Domain perturbation for parabolic quasilinear problems.* Commun. in applied analysis 4 (2000), no. 1, 1–12.

[175] J. SOKOLOWSKI, A. ZOCHOVSKI: *On the topological derivative in shape optimization.* SIAM J. Control Optim. **37** (1999), no. 4, 1251–1272

[176] J. SOKOLOWSKI, J.-P. ZOLESIO: *Introduction to shape optimization. Shape sensitivity analysis.* Springer Series in Computational Mathematics, 16. Springer-Verlag, Berlin (1992).

[177] V. ŠVERÁK: *On optimal shape design.* J. Math. Pures Appl., **72** (1993), 537–551.

[178] L. TARTAR: *Estimations Fines des Coefficients Homogénéises*. Ennio De Giorgi Colloqium, Edited by P.Krée, Res. Notes in Math. **125** Pitman, London, (1985) 168–187.

[179] R. TOADER: *Wave equation in domains with many small obstacles*. Asymptot. Anal. **23** (2000), no. 3-4, 273–290.

[180] N. VAN GOETHEM: *Variational problems on classes of convex domains*. Preprint Dipartimento di Matematica Università di Pisa, Pisa (2000).

[181] A. WAGNER: *A remark on Newton's resistance formula*. Preprint 1998.

[182] H. WEINBERGER: *An isoperimetric inequality for the N-dimensional free membrane problem*. J. Rat. Mech. Anal., **5** (1959), 533–636.

[183] M. WILLEM: *Analyse harmonique réelle*. Hermann, Paris (1995).

[184] S. A. WOLF, J. B. KELLER: *Range of the first two eigenvalues of the Laplacian*. Proc. Roy. Soc. Lond., **A 447** (1994), 397–412.

[185] W.P. ZIEMER: *Weakly Differentiable Functions*. Springer-Verlag, Berlin, 1989.

[186] X. ZHONG: *On nonhomogeneous quasilinear elliptic equations*. Ann. Acad. Sci. Fenn. Math. Diss. No. 117, (1998).

Index

Elenco dei volumi della collana
"Appunti"
pubblicati dall'Anno Accademico 1994/95

GIUSEPPE BERTIN (a cura di), *Seminario di Astrofisica*, 1995.

EDOARDO VESENTINI, *Introduction to continuous semigroups*, 1996.

LUIGI AMBROSIO, *Corso introduttivo alla Teoria Geometrica della Misura ed alle Superfici Minime*, 1997 (ristampa).

CARLO PETRONIO, *A Theorem of Eliashberg and Thurston on Foliations and Contact Structures*, 1997.

MARIO TOSI, *Introduction to Statistical Mechanics and Thermodynamics*, 1997.

MARIO TOSI, *Introduction to the Theory of Many-Body Systems*, 1997.

PAOLO ALUFFI (a cura di), *Quantum cohomology at the Mittag-Leffler Institute*, 1997.

GILBERTO BINI, CORRADO DE CONCINI, MARZIA POLITO, CLAUDIO PROCESI, *On the Work of Givental Relative to Mirror Symmetry*, 1998

GIUSEPPE DA PRATO, *Introduction to differential stochastic equations*, 1998 (seconda edizione)

HERBERT CLEMENS, *Introduction to Hodge Theory*, 1998

HUYÊN PHAM, *Imperfections de Marchés et Méthodes d'Evaluation et Couverture d'Options*, 1998

MARCO MANETTI, *Corso introduttivo alla Geometria Algebrica*, 1998

AA.VV., *Seminari di Geometria Algebrica 1998-1999*, 1999

ALESSANDRA LUNARDI, *Interpolation Theory*, 1999

RENATA SCOGNAMILLO, *Rappresentazioni dei gruppi finiti e loro caratteri*, 1999

SERGIO RODRIGUEZ, *Symmetry in Physics*, 1999

F. STROCCHI, *Symmetry Breaking in Classical Systems and Nonlinear Functional Analysis*, 1999

ANDREA C.G. MENNUCCI, SANJOY K. MITTER, *Probabilità ed informazione*, 2000

LUIGI AMBROSIO, PAOLO TILLI, *Selected Topics on "Analysis in Metric Spaces"*, 2000

SERGEI V. BULANOV, *Lectures on Nonlinear Physics*, 2000

LUCA CIOTTI, *Lectures Notes on Stellar Dynamics*, 2000

Fotocomposizione "CompoMat" Loc. Braccone, 02040 Configni (RI), Italy
Finito di stampare per conto della "CompoMat" dalla Nuova Grafica 86 nel maggio 2002